Inspire Science

Be a Scientist Notebook

Student Journal

Grade 3

Mc
Graw
Hill
Education

mheducation.com/prek-12

STEM McGraw-Hill is committed to providing
instructional materials in Science, Technology,
Engineering, and Mathematics (STEM) that give all
students a solid foundation, one that prepares them
for college and careers in the 21st century.

Send all inquiries to:
McGraw-Hill Education
8787 Orion Place
Columbus, OH 43240

ISBN: 978-0-07-678225-3
MHID: 0-07-678225-5

Printed in the United States of America.

11 12 SWI 25 24 23 22

Our mission is to provide educational resources that enable
students to become the problem solvers of the 21st century
and inspire them to explore careers within Science, Technology,
Engineering, and Mathematics (STEM) related fields.

OWEN
Entomologist

TABLE OF CONTENTS

Check out the activities in every lesson!

OWEN
Entomologist

InspireScience

This is your own personal science journal where you will become scientists and engineers. Use this book to answer questions and solve real-world problems.

This is YOUR journal!
Personalize it!

POPPY
Park Ranger

Motion and Forces

 ## Science in Our World

Susie's school is getting a new playground. Each class was asked to think about how they would like it to look. Susie's dad, an architect, suggested that they make drawings to show how big they want the structures to be and where they could go. Look at the photo of the playground. What questions do you have? What movement do you see? How could these questions help you design a playground?

_____

_____

_____

_____

_____

abc Key Vocabulary

Look and listen for these words as you learn about motion and forces.

accelerate	balanced forces	direction
distance	energy	force
friction	motion	position
speed	unbalanced forces	work

How can I use
what I know about force
and motion to plan
a playground?

SAM
Architectural Drafter

STEM Career Connection
Architectural Drafter

Being an architectural drafter means I am designing all the time. I typically begin with a rough sketch. I will meet with an architect—someone who designs the plans for homes, buildings, or structures—and get an idea of what is needed. This might mean that I have to make detailed plans, use calculations for measurements, or suggest a few different ideas for the architect and client to get the final results they want.

After I make my rough sketch, I will meet with the architect and builder to get their opinions. Once we have met, I take their ideas and continue to improve the design until everybody is satisfied with the drawing. At that point, the "rough draft" has become a blueprint for when construction and building begins.

Why do you think Susie's dad suggested that the class create drawings for the playground design?

_____

_____

_____

Science and
Engineering Practices

I will plan and carry out investigations.

Motion

PAGE KEELEY SCIENCE PROBES

How Far Did the Snail Travel?

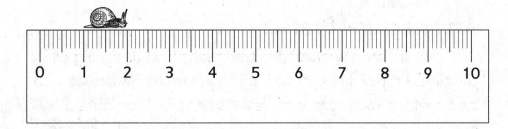

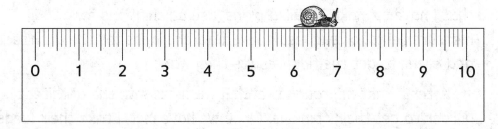

Lupina wondered how far a snail travels in one minute. She placed a snail along a measuring tape as shown in the 1st picture. After one minute, the snail was at the position shown in the 2nd picture. How far did the snail travel?

A. 7 centimeters

B. 6 centimeters

C. 5 centimeters

D. 4 centimeters

Explain your answer. How did you figure out how far the snail moved?

_____

_____

_____

Science in Our World

▶ Watch the video of the boy playing basketball. Observe what is moving. What questions do you have?

_____

_____

_____

_____

Read about a statistician and answer the questions on the next page.

Statisticians use measurements to keep track of information.

STEM Career Connection
Statistician

Collecting data is what I do. Well, it isn't all I do, but it is a huge part of my job. Usually I think about what data or information is needed to solve problems. Then, I figure out how to get that information. Sometimes it is through quick observation. Often I keep track of things over a period of time and see how the numbers change. Once I've collected the data—measurements, scores, or growth figures—I analyze it.

To analyze data means to compare or interpret it. If I keep statistics about a sport, like basketball, I list the baskets that are made by players. Then I analyze the numbers. I could figure how many 3-point shots you made compared to how many you attempted.

CJ
Statistician

1. What does a statistician do after collecting data?

_____

_____

_____

2. Why do you think it is important to interpret the data?

_____

_____

_____

? Essential Question
How can you tell whether something is moving?

_____

_____

_____

⚙ Science and Engineering Practices

I will plan and carry out investigations.

> Like a statistician, you will collect data as you plan and carry out an investigation.

Inquiry Activity
An Object's Position

Materials

☐ classroom object

Can you find an object?

Make a Prediction What words will be used to describe the position of an object?

_____

_____

Carry Out an Investigation

1 Select an object in your classroom.

2 Choose a starting point in the classroom.

3 Write detailed instructions on how to locate the object. Do not identify the object in the description.

_____

_____

_____

_____

4 Trade instructions with a classmate. Observe how your classmate finds the object.

5 **Record Data** Collect your classmate's responses.

What was the object? _____
Were the directions clear to follow?

_____

_____

_____

6 **Analyze Data** Ask your classmate to circle the words in your instructions that best helped them find your object.

Communicate Information

1. Compare the words that your classmate circled to your prediction. Did your results match your prediction? Explain.

_____

_____

_____

2. What could you do to make your instructions better?

_____

_____

_____

3. **Construct an Explanation** How did watching your classmate help you decide how to improve your instructions?

_____

_____

_____

Obtain and Communicate Information

🔤 Vocabulary

Use these words when explaining motion and forces.

position   distance   direction

motion   speed

Position

📖 Read pages 220–221 in the *Science Handbook*. Answer
the following questions after you have finished reading.

1. What words describe position?

_____

_____

2. How could you find the distance from your desk to
the door?

_____

_____

3. Look at page 221. Describe the position of the smaller
bear, using distance and direction.

_____

_____

Things Move

▶ Watch *Things Move* on different types of motion.
Answer the questions after you have finished watching.

4. What is motion?

_____

_____

_____

5. Using your vocabulary words, describe how the girls or boys changed position.

_____

_____

_____

Motion

📖 Read pages 222–223 in the *Science Handbook.* Answer the following question after you have finished reading.

6. How does a straight motion differ from a zigzag motion?

_____

_____

_____

Inquiry Activity
Measure an Object's Speed

You will learn how to determine the speed of
a wind-up toy.

Make a Prediction How fast will the wind-up toy move?

_____

_____

Carry Out an Investigation

1. Lay a meterstick flat on the ground.

2. One partner winds up the toy. Make sure to wind it completely! Place the toy at 0 cm. Do not let go until the other partner says, "Go!"

3. The other partner will use a stopwatch. On "Go!" start the stopwatch and proceed for 30 seconds.

4. Measure the distance the toy traveled. Repeat steps 2–4 for three trials.

5. **Record Data** Write the distance the toy traveled each trial. Double the number to find how far it would travel in 60 seconds (1 minute). Find the speed by dividing the distance by the time.

	Distance Traveled in 30 seconds	Distance Traveled in 1 minute	Speed Distance/Time (cm/minute)
Trial 1	cm	cm	
Trial 2	cm	cm	
Trial 3	cm	cm	

6 **Analyze Data** Using the data collected, create a bar graph on a separate sheet of paper. Show the distance the toy traveled during each trial.

Communicate Information

7. **Construct an Explanation** Why is it important to do more than one trial?

_____

_____

_____

Measuring Motion

📖 Read pages 224–225 in the *Science Handbook.* Answer the following questions after you have finished reading.

8. List the ways motion is described.

_____

_____

9. Review the data you collected for *Measure an Object's Speed*. Use what you read about predicting motion. What would most likely happen in a fourth trial?

_____

_____

Glue your graph here.

FOLDABLES®

Cut out the Notebook Foldables tabs given to you by your teacher. Glue the anchor tabs as shown below. Use what you have learned to describe the movement of the ball using your vocabulary words.

Glue anchor tab here

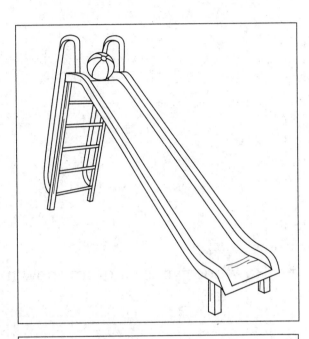

Measure an Object's Speed

10. Revisit the *Measure an Object's Speed* activity on page 11. Draw and label a diagram of your setup. Use lesson vocabulary to label what you observed and measured.

11. How does your diagram show motion?

_____

_____

_____

⚙ Science and Engineering Practices

Use examples from the lesson to explain what you can do!

Think about what you learned about motion. Tell how you can plan and carry out an investigation about motion by completing the "I can . . ." statement below.

I can _____

_____

🔍 Research, Investigate, and Communicate
Moving Through Time

👁 Read *Moving Through Time* on how motion has affected transportation. Answer the following questions after you have finished reading.

1. Why were horses a good choice for travel?

_____   _____

_____

2. How did cars compare to horses?

_____

_____

_____

3. What is the predictable motion of a wheel? How did the predictable motion of wheels enhance transportation?

_____

_____

_____

_____

_____

Performance Task
Motion Models

You will build two models to show patterns of motion. Think about the vocabulary words you have learned in this lesson as you build your models.

Materials

☐ *Science Handbook*

☐ various classroom items

Make a Model

1. Look at the photos on page 223 in the *Science Handbook*.

2. Choose two different patterns of motion to model using classroom items.

3. Make a list of classroom items you will use to make your models.

_____

_____

_____

_____

Like a statistician, observe your model. What data could you collect?

4 Draw your models of the two different patterns of motion below. Label each motion you are modeling.

Name _____ Date _____

Crosscutting Concepts
Patterns

1. Describe a pattern that one of your models shows.

_____

_____

_____

2. Describe how you used a model to show motion. Use your vocabulary words.

_____

_____

_____

_____

3. Pick one of the models. What do you predict about the future motion of that model?

_____

_____

_____

❓ Essential Question
How can you tell whether something is moving?

▷ Think about the video of the boy playing basketball
at the beginning of the lesson. Describe the various types
of motion that you see in the video.

_____

_____

_____

_____

⚙ Science and Engineering Practices

Now that you're done
with the lesson, share
what you did.

Review the "I can . . ." statement you wrote earlier in
the lesson. Explain what you have accomplished
in this lesson by completing the "I did . . ." statement.

I did _____

_____

_____

_____

_____

_____

Forces Can Change Motion

 PAGE KEELEY
SCIENCE
PROBES

Golf Ball

Three friends are playing golf. They each have different ideas about the forces that act on a golf ball. This is what they think:

Finn: *Forces act on the golf ball only when the golfer hits the ball.*

Pete: *Forces act on the golf ball only when the ball is on the tee.*

Tad: *Forces act on the golf ball when it is on the tee and when the golfer hits the ball.*

Who has the best idea about forces? _____

Explain why you think it is the best idea.

_____

_____

_____

_____

Science in Our World

Look at the photo of the building being knocked down or demolished. How is force being used? What questions do you have?

_____

_____

_____

_____

Read about a demolition expert and answer the questions on the next page.

Demolition experts need to understand how forces work when building and demolishing structures.

STEM Career Connection
Demolition Expert

The old Nevada Hotel has been closed and empty for 25 years. The building is dangerous and must be demolished. My plan is to use a crane and wrecking ball to demolish the hotel. There are some new buildings very close to the old hotel. I must use exactly the right amount of force to bring the building down without damaging surrounding buildings. I also must use the wrecking ball in just the right spot to bring the building straight down. If it is used in the wrong place, the force could make the hotel fall over instead of falling straight down.

FINN
Construction Manager

1. Why does the demolition expert need to place the wrecking ball in just the right places?

_____

_____

2. What might happen if the wrecking ball hits in the wrong place?

_____

_____

? Essential Question
How do forces change motion?

_____

_____

Science and Engineering Practices

I will plan and carry out investigations.

> Like a demolition expert, you will learn about how forces work.

Inquiry Activity
Force Affects the Way Objects Move

What is the relationship between force and motion?

Make a Prediction What will happen to an object you are pushing if you increase the force to make it move?

_____

_____

Copyright © McGraw-Hill Education

Materials

- [ ] 6 books
- [ ] cardboard
- [ ] masking tape
- [ ] toy car
- [ ] meterstick

Carry Out an Investigation

1. Stack three books on the floor. Lean a piece of cardboard along the top book to make a ramp. Tape down the edge of the cardboard to the floor.

2. Place a toy car at the top of the ramp. Release the car.

3. **Record Data** Measure and record the distance the car travels.

4. Add three more books to the stack. When you increase the height of the ramp, you increase the force on the toy car. Repeat steps 2 and 3.

Force Affects Objects	
	Distance Traveled
Low Ramp	
High Ramp	

5. **Analyze Data** In the data table, circle the ramp that caused the car to travel the farthest.

Communicate Information

1. What caused the car to move?

_____

_____

2. What was the difference between the two distances traveled?

_____

_____

_____

3. **Construct an Explanation** From your data and observations, how does a force affect the distance an object moves?

_____

_____

💬 Obtain and Communicate Information

🔤 Vocabulary

Use these words when explaining motion and forces.

force	friction	balanced forces
unbalanced forces	accelerate	

Force

▶ Watch *Force* on how pushes and pulls affect motion. Answer the following questions after you have finished watching.

1. Make a list of the types of forces you saw in the video.

_____

_____

Forces and Changes in Motion

📖 Read pages 226–227 in the **Science Handbook.** Answer the following question after you have finished reading.

2. What type of force is needed to move a heavy object?

_____

_____

Changing Forces

🔲 Investigate changing the direction and strength of force by conducting the simulation. Answer the question after you have finished.

3. What happened when you changed the direction of the force?

_____

_____

Types of Forces

📖 Read pages 228–229 in the *Science Handbook.* Answer the following questions after you have finished reading.

4. What is friction?

_____

_____

5. How can you increase friction?

_____

_____

Noncontact Forces

📖 Read pages 230–231 in the *Science Handbook.* Answer the following question after you have finished reading.

6. What does the force of gravity on an object depend on?

_____

_____

_____

Inquiry Activity
Balanced Forces

Materials

☐ book, marker, or chair

You will experiment with changing a force and then balancing it.

Make a Prediction How can you balance a force?

_____

_____

Carry Out an Investigation

1. Place a book, marker, or chair on the floor. Have a partner sit on the opposite side of the object.

2. Have one partner give a slight push on one side of the object.

3. Have the other partner give a slight push on the other side of the object.

4. Both push on the object at the same time with similar force.

5. **Record Data** What happens to the object?

First partner push: _____

Second partner push: _____

Both push: _____

Communicate Information

7. Construct an Explanation Were you able to balance the force on the object? Did your results match your prediction?

_____

_____

_____

Types of Forces—Balanced Forces

📖 Read pages 232–233 in the *Science Handbook.* Answer the following questions after you have finished reading.

8. Describe the forces in balanced forces.

_____

_____

9. Which forces cause a change in motion?

_____

_____

⚙ Crosscutting Concepts
Cause and Effect

10. Revisit the activity Balanced Forces. Reread steps 2 and 4 of the investigation. Label them as Balanced or Unbalanced forces.

FOLDABLES®

Cut out the Notebook Foldables tabs given to you by your teacher. Glue the anchor tabs as shown below. Use what you have learned throughout the lesson to describe the picture using vocabulary words.

Glue anchor tab here

Types of Forces

📖 Read pages 234–235 in the *Science Handbook*.
Answer the following questions after you have
finished reading.

11. What does it mean to *accelerate*?

_____

_____

12. Which of the two bicycles might accelerate more
rapidly—a big, old, heavy bike with a basket filled with
stuff or a more thin, light one?

_____

_____

⚙ Crosscutting Concepts
Cause and Effect

13. On page 235 in the *Science Handbook,* the third
drawing shows that the load accelerates half as fast as
in the first drawing. What causes this?

_____

_____

Changing Forces

Investigate how changing the direction and strength of a force affects acceleration by revisiting the simulation. Answer the questions after you have finished.

14. What happens to the acceleration of the puck as you increase the strength of the force?

_____

_____

15. How are you able to completely stop the puck?

_____

_____

_____

Science and Engineering Practices

Use examples from the lesson to explain what you can do!

Think about how forces can change motion. Tell how you can plan and carry out an investigation by completing the "I can . . ." statement below.

I can _____

_____

_____

Research, Investigate, and Communicate

Inquiry Activity
Friction Affects Force

You will learn about how friction affects force.

Make a Prediction How does friction affect the force needed to move an object?

_____

_____

Materials
☐ 6 books
☐ cardboard
☐ masking tape
☐ toy car
☐ meterstick
☐ sandpaper
☐ cotton towel

Carry Out an Investigation

1. Re-create the *Force Affects the Way Objects Move* activity with 6 books (high ramp). Use the data from that investigation.

2. Tape a layer of sandpaper at the bottom of the ramp. Release the car from the top of the ramp.

3. **Record Data** Measure and record the distance the car travels.

4. Remove the sandpaper. Tape a cotton towel at the bottom of the ramp. Release the car from the top of the ramp.

5. **Record Data** Measure and record the distance the car travels.

Friction Affects Force	
	Distance Traveled
High Ramp	
High Ramp with sandpaper	
High Ramp with cotton towel	

6 **Analyze Data** Use the data collected to create a graph showing the relationship between the material on the ramp and the distance the car traveled.

Communicate Information

1. Which material created more friction with the car, the towel or the sandpaper? Explain.

_____

_____

_____

2. **Construct an Explanation** Does your data support your prediction? Why or why not?

_____

_____

_____

_____

_____

Glue your graph here.

⚙ Performance Task
Building Demolition

You will use force to destroy a model building. The demolition experts would like you to plan how to knock over the model building. They need 4 cups to stay standing after the demo is done.

Materials

☐ 9 plastic cups

☐ 1 rubber ball

Write a Hypothesis If I _____

_____

then _____

_____

_____

Carry Out an Investigation

1. On a separate piece of paper, list your procedure.

2. **Record Data** What are your results?

_____

_____

Communicate Information

1. Summarize your investigation and findings.

_____

_____

_____

Glue your procedures here.

2. Construct an Explanation Do your observations and data support your hypothesis?

_____

_____

_____

? Essential Question

How do forces change motion?

Think about the photo of the building demolition at the beginning of the lesson. How did force change the motion of the building?

_____

_____

_____

_____

⚙ Science and Engineering Practices

Now that you're done with the lesson, share what you did!

Review the "I can . . ." statement you wrote earlier in the lesson. Explain what you have accomplished in this lesson by completing your "I did . . ." statement.

I did _____

_____

_____

_____

Simple Machines

PAGE KEELEY
SCIENCE
PROBES

Is It a Simple Machine?

Playing on a seesaw	Using a motor to make a boat go fast	Pushing a box up a ramp
Removing a bottle cap with a bottle opener	Stirring soup with a spoon	Lifting a bucket from a well with a pulley
Pulling a suitcase that has wheels	Cutting paper with scissors	Baking cookies in an oven

People use different kinds of simple machines every day.
Put an X in the boxes that are examples of using
a simple machine.

Explain your thinking. How did you decide if something is a simple

machine? _____

_____

Science in Our World

Look at the photos of the crane and the pulley. What can these machines be used for? What questions do you have?

_____

_____

_____

Read about a construction manager and answer the questions on the next page.

> Like a construction manager, you will learn how machines help us work.

STEM Career Connection
Construction Manager

    I'm heading out to a job today. We are building a multi-story office building near downtown. At this point in a project, you can see the actual building in progress. Just a few months ago, the building was just a blueprint on a piece of paper. When I got that blueprint, I planned a schedule for the project, including deadlines. Then I figured out how many people we would need to get the work done. Not only that, but I had to make sure we had all of the machines and tools we needed. As we have been building, I am observing to make sure that we are following the rules and safety codes. Just yesterday I had to apply for a permit to make sure our electricity is done the right way. I communicate along the way with architects, clients, suppliers of materials, and anyone else who is a part of the project.

FINN
Construction Manager

1. What did the construction manager do after receiving the blueprints?

_____

_____

_____

2. Other than people, what else did the construction manager need to think about to get the work done?

_____

_____

_____

? Essential Question
How do simple machines make work easier?

_____

_____

_____

⚙ Science and Engineering Practices

I will plan and carry out investigations.

Just like a construction manager, you will plan and carry out an investigation to understand simple machines.

Inquiry Activity
Simple Machines Lift Objects

How do simple machines help us lift objects?

Make a Prediction How will moving the ruler's position on a marker change the amount of force needed to lift the blocks?

_____

_____

_____

Copyright © McGraw-Hill Education

Carry Out an Investigation

1. Use clay to attach a marker to the center of a ruler.

2. Use clay to attach a small cup to the top of each end of the ruler.

3. Place two large blocks into one cup.

4. Add gram cubes to the other cup. How many cubes does it take to lift the larger blocks?

5. **Record Data** Record the number of gram cubes in the table.

6. What happens if you change a variable in the experiment? Change the positon of the marker. Move it closer to the end of the ruler with the gram cubes.

7. Repeat steps 3–5.

8. Change the same variable again. Move it closer to the other end of the ruler.

9. Repeat steps 3–5.

Materials

☐ clay

☐ thick marker or wooden dowel

☐ ruler

☐ 2 small plastic drinking cups

☐ 2 large blocks

☐ Several 1-gram cubes

10 **Analyze Data** Circle the position of the marker that required more force to lift the large blocks.

Force Needed to Lift Blocks	
Position of Marker	**Number of Gram Cubes**
Center of Ruler	
Closer to the Gram Cube Cup	
Closer to the Block Cup	

Communicate Information

1. In the activity, what represents the force needed to lift the blocks?

_____

_____

2. What happened when the marker was closer to the block cup?

_____

_____

3. **Construct an Explanation** How did your observations compare to your prediction?

_____

_____

_____

Obtain and Communicate Information

abc Vocabulary

Use these words when explaining simple machines.

work	simple machines	load
wheel and axle	inclined plane	compound machine

Work and Simple Machines

▶ Watch *Work and Simple Machines* on the many machines that help us do work. Answer the question after you have finished watching.

1. Draw an example of work that was done in the video.

Work and Machines

📖 Read page 236 in the *Science Handbook.* Answer the following questions after you have finished reading.

2. What is work?

_____

_____

_____

3. Try each action listed in the table. Is it work? Explain.

Work		
Action	Is it Work?	Why or Why Not
Pick up a book		
Think about a problem		
Slide a chair		
Press feet against the floor		
Push against a wall		

4. What does a machine do? What does a machine not do?

_____

_____

_____

Types of Simple Machines

📖 Read pages 237–242 in the *Science Handbook*. Answer the following questions after you have finished reading.

5. What are simple machines?

_____

_____

6. Fill in the table using information from the *Science Handbook* or other sources. Tell about each simple machine. Draw or write an example of each.

Lever	Pulley	Wheel and axle
Inclined plane	Screw	Wedge

7. Select one of the simple machines from the table.
_____ Look at the example. Explain how this machine makes work easier.

_____

_____

_____

FOLDABLES

Cut out the Notebook Foldables tabs given to you by your teacher. Glue the anchor tabs as shown below. Describe two machines you have used that have no moving parts and two that have few moving parts.

Glue anchor tab here

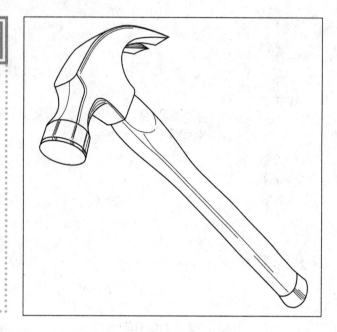

Glue anchor tab here

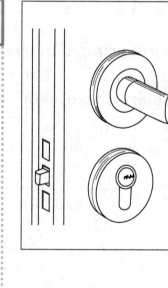

Machines Working Together

Read page 243 in the *Science Handbook.* Answer
the following questions after you have finished reading.

8. How are compound and simple machines related?

_____

_____

_____

9. Draw a picture of a compound machine that you
have used.

Science and Engineering Practices

Think about how you have conducted
investigations about simple machines. Tell
how you can plan and carry out investigations
by completing the "I can . . ." statement
below.

I can _____

_____

_____

*Use examples from
the lesson to explain
what you can do!*

Research, Investigate, and Communicate

Inquiry Activity
Build a Simple Machine

You will investigate whether a simple machine can make work easier.

Make a Prediction Does a pulley make work easier?

_____

_____

_____

Carry Out an Investigation

1. Lift the soda bottle with one hand.

2. Make a pulley. Slide the dowel through the spool. Place the dowel so that each end is on a different desk or table surface. Tape the dowel to the desks.

3. Tie one end of the ribbon to the soda bottle.

4. Place the ribbon over the top of the spool. Pull the ribbon. Notice how the soda bottle is lifted.

5. Record data Lift the soda bottle without the pulley and with the pulley. Measure the force with a spring scale.

Without Pulley	With Pulley

Communicate Information

1. Draw the simple machine. Label the parts. Identify the load.

2. **Construct an Explanation** Did your observations match your prediction? How did the machine make work easier?

_____

_____

_____

Crosscutting Concepts
Cause and Effect

3. Describe the effect of using a simple machine.

_____

_____

_____

Performance Task
Test a Simple Machine

You will use a lever to lift an object and measure the force required.

Make a Prediction How will the needed force change as you move the fulcrum of the lever?

_____

_____

_____

Carry Out an Investigation

1. List your procedure.

_____

_____

_____

2. **Record Data** Create a table that shows your data.

Materials	
☐	thick marker or wooden dowel
☐	ruler
☐	2 small plastic drinking cups
☐	several 1-gram cubes
☐	item to lift

Communicate Information

1. Construct an Explanation Did your data confirm your prediction?

_____

_____

_____

? Essential Question
How do simple machines make work easier?

Think about the photos of the crane and pulley you saw at the beginning of the lesson. How do the crane and pulley help make work easier?

_____

_____

_____

Now that you are done with the lesson, share what you did!.

⚙ Science and Engineering Practices

Review the "I can . . ." statement you wrote earlier in the lesson. Explain what you have accomplished in this lesson by completing the "I did . . ." statement.

I did _____

_____

_____

Motion and Forces

⚙ Performance Project
Observe Motion on a Playground

Take a walk on your playground, or use the Internet to find videos of playgrounds in action.

Describe the motions you observe. Tell how many times something happened, or how much motion or force was used. Notice the distance between different activities. Ask yourself why the distances are important.

_____

_____

_____

_____

On the next page, draw a model of an ideal playground. Use the information you gathered to help make your decisions. Label your drawings to show what you know. What forces and motions are you including?

Use what you've learned about force and motion to design a playground.

🌎 Explore More in Our World

Did you learn the answers to all of your questions from the beginning of the module? If not, how could you design an experiment or conduct research to help answer them?

Electric and Magnetic Forces

 Science in Our World

The gate in the fence around the garden will not stay shut.
The farmers decide to use a magnetic gate. What questions
do you have about a magnetic gate?

_____

_____

_____

_____

abc Key Vocabulary

Look and listen for these words as you learn
about electric and magnetic forces.

attract

magnetic field

repel

electrical
charge

magnetism

magnet

pole

static electricity

How might electric and magnetic forces help a gate stay closed?

HANNAH
Welder

STEM Career Connection
Electrical Engineer

As an electrical engineer, I evaluate electrical systems and products, and I use what I know to help design and research how to use them. Sometimes I have to test the items to see what is working and what is not working, and try to figure out why. It is important that I keep good notes about what I am finding. I write down data from my research and graph it too. That way I can see if there is a trend, and whether this will help me solve the problem.

An electrical engineer can work on anything related to electricity, magnetism, and electronics. I work on many products—from computers, to robots, to cell phones, to wiring in a building.

Draw and label a diagram to show how you think electricity and magnets might help a gate stay closed.

Science and Engineering Practices

I will ask questions and define problems.

Electricity

**PAGE KEELEY
SCIENCE
PROBES**

Salma's Hair

Salma rubbed a balloon on her hair. She then held the balloon over her head. Salma and her friends laughed when her hair went straight up and stuck to the balloon. They each had different ideas about why Salma's hair went up toward the balloon. Here is what they said:

Salma: *I think the balloon is acting like a magnet on my hair.*

Curt: *I think the balloon and hair are electrically charged.*

Nicki: *I think the balloon is taking energy from the hair.*

Who do you agree with most? _____

Explain why you agree.

_____

_____

_____

_____

_____

Science in Our World

Look at the photo of the balloon and the girl's hair. What is happening to her hair? What questions do you have?

_____

_____

_____

_____

Read about an electronics technician and answer the questions on the next page.

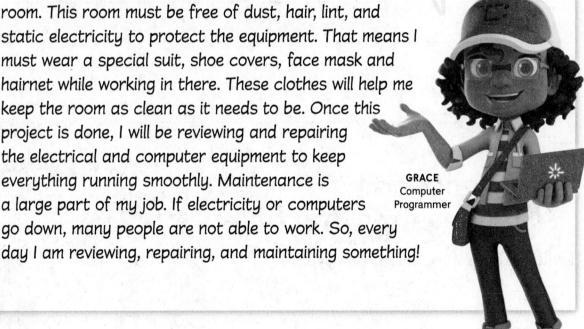

Electronics technicians need to understand how electricity and magnets can solve problems.

STEM Career Connection
Electronics Technician

Today we are building new computers in a "clean" room. This room must be free of dust, hair, lint, and static electricity to protect the equipment. That means I must wear a special suit, shoe covers, face mask and hairnet while working in there. These clothes will help me keep the room as clean as it needs to be. Once this project is done, I will be reviewing and repairing the electrical and computer equipment to keep everything running smoothly. Maintenance is a large part of my job. If electricity or computers go down, many people are not able to work. So, every day I am reviewing, repairing, and maintaining something!

GRACE
Computer
Programmer

1. Why must the room be free of dust, hair, lint and static elecricity?

_____

_____

_____

2. Why is maintenance such a large part of the electronics technician's job?

_____

_____

_____

? Essential Question
How does electric force affect objects?

_____

_____

_____

Science and Engineering Practices

I will ask questions and define problems.

Like an electronics technician, you will ask questions and define problems.

Inquiry Activity
Charged or Uncharged Balloons

How can you observe balanced or unbalanced forces?

Make a Prediction What occurs when charges between two objects are unbalanced?

_____

_____

_____

Carry Out an Investigation

1. Label the balloons A, B, C, D, and E. Blow up each balloon and tie it.

2. Balloon A is the control sample. Hold the balloon close to the confetti without touching it.

3. **Record Data** Record your observations in the table after each step.

4. Balloon B is the cotton sample. Rub the balloon with the cotton cloth for 20 seconds. Hold the balloon close to the confetti without touching it.

5. Balloon C is the paper towel sample. Rub the balloon with the paper towel for 20 seconds. Hold the balloon close to the confetti without touching it.

6. Balloon D is the acrylic sample. Rub the balloon with the acrylic sock for 2 minutes. Hold the balloon close to the confetti without touching it.

7. Balloon E is the wool sample. Rub the balloon with the wool fabric for 2 minutes. Hold the balloon close to the confetti without touching it.

Materials

- [ ] permanent marker
- [ ] 5 balloons
- [ ] paper plate
- [ ] 1/4 cup paper confetti (hole-punched paper circles)
- [ ] piece of cotton cloth
- [ ] paper towel
- [ ] acrylic sock
- [ ] piece of wool (sock or sweater)

| Material's Effect on Balloon ||
Sample	Observations
Sample A—Control	
Sample B—Cotton	
Sample C—Paper Towel	
Sample D—Acrylic	
Sample E—Wool	

8 **Analyze Data** Circle the sample that had the largest charge. Underline the sample that had the smallest charge.

Communicate Information

1. Describe the sample that had the largest charge.

_____

_____

_____

2. **Construct an Explanation** How did your observations compare with your prediction?

_____

_____

_____

Obtain and Communicate Information

Vocabulary

Use these words when explaining electricity.

**static electricity   electrical charge attract
repel**

Static Electricity

▶ Watch *Static Electricity* on how static electricity can affect objects. Answer the question after you have finished watching.

Crosscutting Concepts
Cause and Effect

1. What example of static electricity was shown in the video? What effect did the static electricity have on the object?

_____

_____

Electrical energy

📖 Read pages 254–255 in the *Science Handbook.* Answer the following questions after you have finished reading.

2. What causes electricity?

_____

_____

3. Draw and label a diagram that shows the different combinations of electrical charges and how they interact with each other.

Positive and Negative Charges

Explore the Digital Interactive *Positive and Negative Charges* on static electricity. Answer the questions after you have finished.

4. Use vocabulary words to summarize what happened to the balloon.

_____

_____

_____

_____

5. How does this relate to the Inquiry Activity *Charged or Uncharged Balloons* that you completed?

_____

_____

_____

_____

_____

_____

6. What questions do you have about static electricity and the distance of objects?

_____

_____

_____

Does Distance Make a Difference?

Investigate creating static electricity charges on a balloon with different materials by conducting the simulation. Answer the questions after you have finished.

7. Record Data Create a table showing what happens with different materials in *Does Distance Make a Difference?*

Does Distance Make a Difference?			
Item	Number of Rubs	Distance from Paper	Observations

8. What conclusions can you draw?

_____

_____

_____

FOLDABLES

Cut out the Notebook Foldables given to you by your
teacher. Glue the anchor tabs as shown below.
Describe the picture using vocabulary words.

Glue anchor tab here.

Inquiry Activity
Eliminate Static Electricity

You will investigate which, if any, dryer sheets help eliminate static electricity.

Define a Problem What problem does static electricity cause?

_____

_____

_____

Ask a Question What question do you hope to answer from the investigation?

_____

_____

_____

Carry Out an Investigation

1. Write a procedure to help answer the question.

_____

_____

2. **Record Data** Create a table to share your findings.

Materials

☐ 1 balloon

☐ paper plate

☐ paper confetti (hole-punched paper circles)

☐ a piece of wool (sock or sweater)

☐ various dryer sheets

Online Content at connectED.mcgraw-hill.com

Communicate Information

9. **Construct an Explanation** What conclusions can you draw from the data?

_____

_____

_____

10. If you were to purchase a dryer sheet to reduce static electricity, which would you buy?

_____

_____

_____

Science and Engineering Practices

Use examples from the lesson to explain what you can do.

Think about how you asked questions to learn more about static electricity. Tell how you can ask questions by completing the "I can . . ." statement below.

I can _____

_____

_____

🔍 Research, Investigate, and Communicate

Electricity in Nature

👁 Read **Electricity in Nature** on the forms of electricity that occur in nature. Answer the questions after you have finished reading.

1. How are lightning and a static electricity spark similar?

_____

_____

2. Describe how animals use electricity.

_____

_____

_____

✋ Inquiry Activity
Make Lightning

You will make sparks of "lightning" using static electricity.

Make a Prediction How will you make sparks?

_____

_____

_____

Carry Out an Investigation

1. Cut a piece off one corner of the tray. Tape it to the center of the pie tin to make a handle.

2. Quickly rub the bottom of the polystyrene tray on your hair.

3. Put the tray upside down on a table.

4. Use the handle to pick up the pie tin.

5. Drop the tin on top of the tray.

6. Very slowly touch the tip of your finger to the pie tin.

7. Use the handle to pick up the pie tin again. Touch the pie tin with your finger.

8. Repeat steps 4–7 again. If the sparks stop, then start at step 2.

Materials

- ☐ scissors
- ☐ polystyrene tray
- ☐ masking tape
- ☐ aluminum pie tin

Communicate Information

1. Summarize what you observed.

_____

_____

2. Draw what you saw.

⚙ Performance Task
Teach Static Electricity

You will create static electricity and explain what is happening at each step.

Ask a Question What question could you ask to help prepare for your task?

_____

_____

_____

Carry Out an Investigation

① Plan the investigation.

_____

_____

_____

_____

Materials

☐ balloon

☐ paper plate

☐ polystyrene tray

☐ 1/4 cup paper confetti (hole-punched paper circles)

☐ piece of cotton cloth

☐ acrylic sock

☐ piece of wool (sock or sweater)

Remember that an electronics technician asks questions to solve problems.

Communicate Information

1. Show what you learned. Draw and label a diagram of your investigation to tell what you've learned about electrical charges and static electricity.

❓ Essential Question
How does electric force affect objects?

Think about the photo of the balloon at the beginning of the lesson. How did electric force affect the balloon?

_____

_____

_____

⚙️ Science and Engineering Practices

Review the "I can . . ." statement you wrote earlier in the lesson. Explain what you have accomplished in this lesson by completing the "I did . . ." statement.

Now that you're done with the lesson, share what you did!

I did _____

_____

_____

_____

_____

_____

Magnets

Magnet and Paper Clip

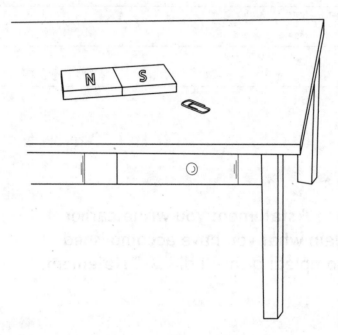

Magnets are used to attract objects. They move toward the magnet. Magnets are also used to repel objects. They move away from the magnet. What happens when a bar magnet is placed near a steel paper clip? Circle the best answer.

A. Both ends of the magnet will attract the paper clip.

B. One end of the magnet attracts the paper clip. The other end repels the paper clip.

C. Both ends of the magnet repel the paper clip.

Explain your thinking. How did you decide what happens with both ends of the magnet?

_____

_____

Science in Our World

Look at the photo of magnets levitating on the pencil. What questions do you have?

_____

_____

_____

_____

Read about a magician and answer the questions on the next page.

STEM Career Connection
Magician

I would like to create a new magic trick for my show. My audiences love tricks that make objects float or levitate. Some of my tricks involve the use of magnets. Magnets can push or pull each other. Even though the tricks look magical, they are really just science demonstrations. One trick that I do for birthday parties involves making an object levitate on a pencil. Children are amazed by this trick. I always get loud applause.

Magicians use magnets to entertain us!

ANTONIO
Robotics Engineer

1. How does the magician use magnets?

_____

_____

_____

2. Instead of doing a magic trick, what is the magician really doing with magnets?

_____

_____

_____

? Essential Question
How can you use a magnet?

_____

_____

_____

Science and Engineering Practices

I will ask questions and define problems.

Like a magician you will ask questions and define problems that magnets can solve.

Inquiry Activity
Investigate with Magnets

What effect do magnets have on different objects?

Make a Prediction Which objects will be pulled to a magnet?

_____

_____

_____

Materials
☐ magnet
☐ various classroom objects
☐ pennies
☐ paper clips
☐ pencils
☐ crayons
☐ plastic spoons

Carry Out an Investigation

BE CAREFUL: Don't place science materials in your mouth.

1. Lay out the items you are going to test on a table.

2. Test each item by touching it with the magnet.

3. **Record Data** List each item in the table. Tell if it was pulled to the magnet.

Item	Pulled or Not Pulled by Magnet

Communicate Information

1. Analyze Data Describe the findings from your data table.

_____

_____

_____

_____

_____

2. Construct an Explanation How did your observations compare to your prediction?

_____

_____

_____

_____

_____

_____

💬 Obtain and Communicate Information

🔤 Vocabulary

Use these words when explaining magnets.

magnet   **magnetic field**   **pole**

magnetism

Magnets

📖 Read pages 259–260 in the *Science Handbook*. Answer the following questions after you have finished reading.

1. What is magnetism?

_____

_____

2. Draw how two bar magnets can attract and repel each other. Use N and S to label the poles.

Exploring Magnets

▶ Watch *Exploring Magnets* to learn more about magnets. Answer the question after you have finished watching.

3. Did any of the information from the video confirm your findings in the Investigate with Magnets activity?

_____

_____

✋ Inquiry Activity
Distance and the Pull of a Magnet

Materials

☐ ruler

☐ 2 bar magnets

You will measure the distance between magnets and observe the attraction.

Write a Hypothesis If _____

then _____

because _____

Carry Out an Investigation

① Place the ruler on a flat table or desktop.

② Put the north end of one bar magnet and the south end of another magnet at the 15 centimeter mark of the ruler.

③ **Record Data** Create a table to record your observations.

④ Leave the bar magnet with the north end at the 15 centimeter mark. Move the other magnet to the 17 centimeter mark. Repeat Step 3. Try another distance.

Glue your table here.

Communicate Information

4. Summarize your observations.

_____

_____

_____

5. **Construct an Explanation** Was your hypothesis supported
   by your observations?

_____

_____

_____

Magnetic Field

📖 Read pages 261–262 in the *Science Handbook.* Answer
   the following questions after you have finished reading.

6. How did the teacher demonstration show you
   the magnetic field?

_____

_____

_____

7. Why does a compass arrow always point north?

_____

_____

_____

Inquiry Activity
Magnetic Forces Pass Through Objects

You will investigate if magnetic force passes through or is blocked by objects.

Ask a Question Write a question that you would like to answer in your investigation.

_____

_____

Materials
☐ aluminum foil sheet
☐ plastic plate
☐ 1 sheet of notebook paper
☐ textbook
☐ wooden desktop or tabletop
☐ 2 ring magnets

Carry Out an Investigation

1 **Test** each object. Place the magnets so they will attract one another on either side of the objects to be tested.

2 **Record Data** How do the magnets react? Record the reaction in the table.

Object	Did magnetic force pass through?
aluminum foil sheet	
plastic plate	
1 sheet of notebook paper	
textbook	
wooden desktop or tabletop	

Communicate Information

1. **Construct an Explanation** Was your question answered?

_____

_____

FOLDABLES

Cut out the Notebook Foldables given to you by
your teacher. Glue the anchor tabs as shown below.
Describe the picture using vocabulary words.

Glue anchor tab here.

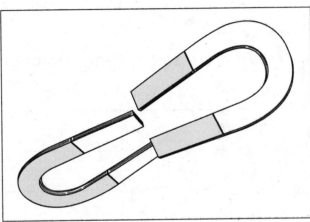

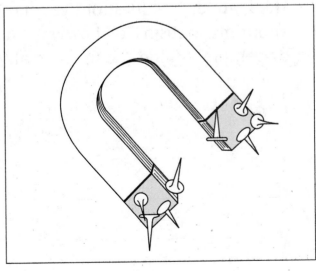

Exploring Magnets

 Explore the Digital Interactive *Magnets* on how they push and pull. Answer the questions after you have finished.

8. How are magnets used on a refrigerator?

_____

_____

_____

9. What are some other uses of magnets?

_____

_____

_____

Science and Engineering Practices

Use examples from the lesson to explain what you can do!

Think about the questions you have asked about magnetism. Tell how you can ask a question by completing the "I can ..." statement below.

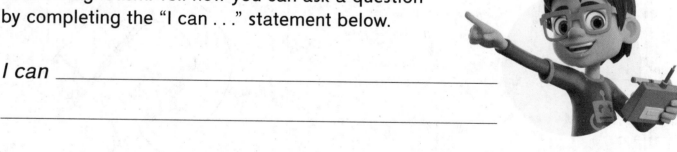

I can _____

_____

_____

Research, Investigate, and Communicate

Inquiry Activity
Make an Electromagnet

You will construct an electromagnet and consider how to make it stronger.

Write a Hypothesis If _____

_____

then _____

_____ because _____

_____

Materials

☐ D-cell battery

☐ 1 iron nail

☐ 1 battery holder

☐ 1 40 -centimeter long piece of insulated wire

☐ paper clips

Carry Out an Investigation

BE CAREFUL Wire may become warm in this activity.

1. Wind 40 centimeters of insulated wire around an iron nail 20 times. Start at one end of the nail with the insulated wire. Leave at least 4 centimeters of wire at the starting end.

2. Place the D-cell battery into the battery holder.

3. Attach the ends of the insulated wire into the clips on each end of the battery holder.

4. Use the nail as a magnet and see how many paper clips you can pick up. Record the results in the table.

5. Repeat the procedure. Wind the wire 10 more times around the nail.

6. Repeat the procedure, winding the wire 10 more times.

7 Record Data

Strength of Electromagnet	
Number of Times Wire Is Wound Around Nail	Number of Paper Clips
20	
30	
40	

Communicate Information
Crosscutting Concepts
Cause and Effect

1. How did the number of times the wire was wound affect the number of paper clips picked up?

_____

_____

_____

2. **Construct an Explanation** Was your hypothesis proved?

_____

_____

_____

⚙ Performance Task
Become a Levitation Magician

Use what you know about magnets to re-create the levitation trick.

Ask a Question What will you need to consider as you re-create the trick?

_____

_____

_____

_____

Carry Out an Investigation

1 Look at the photo of the levitation trick. Write the steps it will take to re-create the trick.

_____

_____

_____

_____

_____

Use what you know about magnets to become a magician!

Communicate Information

1. **Construct an Explanation** Were you able to re-create the magic trick? Use evidence to explain.

_____

_____

_____

2. Draw a model of the levitation trick. Label the parts and North and South poles. Use vocabulary words from this module.

❓ Essential Question
How can you use a magnet?

Think about the photo of the magnets levitating on the pencil at the beginning of the lesson. How are the magnets used to create this trick?

_____

_____

_____

_____

⚙️ Science and Engineering Practices

Now that you're done with the lesson, share what you did!

Review the "I can . . ." statement you wrote earlier in the lesson. Explain what you have accomplished in this lesson by completing the "I did . . ." statement.

I did _____

_____

_____

_____

_____

Electric and Magnetic Forces

⚙️ Performance Project
Solve a Simple Design Problem

The farmer needs your help! Create a design solution that will keep the gate around his garden shut.

Make a list of ways you could solve the problem.

_____

_____

_____

_____

_____

Do any of your solutions use magnets? If not, how could you incorporate magnets?

_____

_____

_____

_____

Use what you learned about magnetism and electricity to design a solution!

Design a sketch of the solution to the gate problem using magnets. Label the sketch.

How does the design solve the gate problem?

_____

_____

_____

How could you improve the solution?

_____

_____

 ## Explore More in Our World

Did you learn the answers to all of your questions from the beginning of the module? If not, how could you design an experiment or conduct research to help answer them?

Weather and Climate

 ## Science in Our World

Study the picture of the weather map. What do you observe in the picture? What questions do you have about the picture?

_____

_____

_____

_____

_____

_____

abc Key Vocabulary

Look and listen for these words as you learn about weather and climate.

air pressure	atmosphere	axis
climate	cloud	precipitation
season	weather	wind

What information does a meteorologist need to predict the weather?

POPPY
Park Ranger

STEM Career Connection
Broadcast Meteorologist

July 21-Tuesday

The current weather conditions are a temperature of 23° C (73° F), partly cloudy, with a 40% chance of thunderstorms this evening. This weekend will be hot with a high temperature of 37° C (99° F) and a low of 24° C (75° F). I also see no rain for the weekend. So, now is the time to plan those cookouts and pool parties! Based on weather models from the National Weather Service, it looks as though we will have an early, cool, and wet autumn. Temperatures will be great for those evening football games!

What questions do you have for a meteorologist?

Science and Engineering Practices

I will analyze and interpret data.

I will obtain, evaluate, and communicate information.

Weather Changes

What Happened to the Puddle?

Four friends were walking to school. They noticed a big puddle on the sidewalk. An hour later, the puddle was gone. They wondered where the water in the puddle went. This is what they said:

Henry: *It just dried up and does not exist anymore.*

Sofia: *It went up in the sky and is now part of a cloud.*

Peyton: *It went into the air around us.*

Micah: *It went into a river or a lake.*

Which friend do you agree with the most? _____

Explain why you agree.

_____

_____

Science in Our World

Look at the photos of extreme weather.
What questions do you have?

_____

_____

_____

_____

_____

Read about a meteorologist and answer
the questions on the next page.

> Meteorologists analyze weather data so they can predict what might happen next.

STEM Career Connection
Meteorologist

I am on my way to the office because
we have a huge storm coming. Once I get there,
I will be able to use many tools to study the
meteorological data that has been gathered from
our surface stations, in the upper-air stations, and
from satellites and radar. I will analyze and interpret
these data to help prepare forecasts, which are reports
that tell what is likely going to happen next. Some of
these tools measure and record air pressure every five
minutes. I use all of those data to create charts and
maps that help me predict what is going to happen in
the next few hours as well as in the next week or two.

When extreme weather is happening, I have busy days.
I study the charts, provide reports, and make sure I am
giving the best information that I can. That way, people
can prepare and be safe.

HUGO
Meteorologist

1. Why do meteorologists analyze and interpret data?

_____

_____

_____

2. Why is it important for a meteorologist to provide good information?

_____

_____

_____

? Essential Question
How does weather change?

_____

_____

_____

Science and Engineering Practices

I will analyze and interpret data.
I will obtain, evaluate, and communicate information.

Like a meteorologist, you will analyze and interpret data and communicate information.

Inquiry Activity
Air Is Around You

How can you confirm the presence of air?

Make a Prediction Can air keep a paper towel inside a cup from becoming wet?

_____

_____

_____

Materials

- [ ] plastic container
- [ ] water
- [ ] paper towel
- [ ] plastic cup

Carry Out an Investigation

1. Fill a container two-thirds full with water. Stuff a dry paper towel in the bottom of the cup.

2. Hold the cup upside down over the water. Push the cup straight down into the bottom of the container. Do not tilt the cup.

3. Lift the cup straight out of the water. Do not tilt it. How does the paper towel feel?

_____

_____

4. Repeat step 2. Slowly tilt the cup. What do you observe?

_____

_____

_____

Communicate Information

1. What escaped from the cup in step 4?

_____

_____

_____

2. **Construct an Explanation** How did your observations compare with your prediction?

_____

_____

3. Draw and label your activity when the cup was in the water. Show where the air was.

Obtain and Communicate Information

🔤 Vocabulary

Use these words when explaining weather changes.

weather	atmosphere	precipitation
air pressure	wind	cloud

Weather

📖 Read pages 152–153 in the *Science Handbook.* Answer the following questions after you have finished reading.

1. Where does weather occur?

_____

_____

2. What causes the air temperature to change?

_____

_____

Weather Report

▶ Watch *Weather Report* on predicting weather. Answer the question after you have finished watching.

3. According to the video, a meteorologist studies weather patterns. What does this mean?

_____

_____

Weather—Describing and Measuring Weather

📖 Read pages 154–155 in the *Science Handbook.* Answer
the following questions after you have finished reading.

4. What is precipitation?

_____

_____

_____

5. What is the relationship between wind and air pressure?

_____

_____

_____

Read a Weather Report

When you interpret data, you use the information that has
been gathered to answer questions or to solve problems.
It is easier to interpret data when it is shown in a table or
a graph.

6. Use the data table to make a line graph.

Average Monthly Air Temperature in Atlanta, Georgia (C°)											
Jan.	Feb.	Mar.	Apr.	May	June	July	Aug.	Sept.	Oct.	Nov.	Dec.
5	7	12	16	21	24	26	26	23	17	18	7

7. Analyze the data in the table and in your line graph.
Which months are coolest? Which months are warmest?

_____

_____

_____

Glue your graph here.

Predicting Weather

📖 Read pages 156–157 in the *Science Handbook*. Answer the following questions after you have finished reading.

8. What do meteorologists use tools for?

_____

_____

_____

_____

Glue your graph here.

✋ Inquiry Activity
Predict Weather

You will use tools to predict the weather where you live.

Ask a Question What question will your research help to answer?

_____

_____

Materials
☐ weekly weather report
☐ current weather map

1. Look at a current weekly weather report. Create a bar graph that shows the high and low temperature for each day.

2. Look at a current weather map. Look to the west of where you live. Look to the east.

Communicate Information

9. Is there a relationship between the weather to our east, and our weather yesterday?

_____

_____

_____

10. **Construct an Explanation** Were you able to answer your question?

_____

_____

_____

The Water Cycle

Explore the Digital Interactive *The Water Cycle* on how water affects weather changes. Answer the following questions after you have finished.

11. What happens when water condenses?

_____

_____

12. What does water become when it evaporates?

_____

FOLDABLES®

Cut out the Notebook Foldables given to you by
your teacher. Glue the anchor tabs as shown below.
Describe the stages of the water cycle.

Glue anchor tab here

Clouds

📖 Read pages 158–159 in the *Science Handbook.* Answer the following questions after you have finished reading.

13. Tell about the information you can learn from clouds.

_____

_____

Science and Engineering Practices

Think about how you have analyzed, interpreted and communicated information about weather changes. Tell how you can analyze and interpret data as well as obtain, evaluate, and communicate information by completing the "I can . . ." statement below.

Use examples from the lesson to explain what you can do!

I can _____

_____

_____

_____

🔍 Research, Investigate, and Communicate
Write a Script

📖 Read pages 168–173 in the *Science Handbook* and other resources that provide information about weather events. Work with a group to write a script.

🖱 Explore the Digital Interactive *Weather Events* on types of extreme weather. Write a description of each photo as though you are a broadcast meteorologist. Use this table to organize your information.

Photo	Type of Weather	Details to Include
1		
2		
3		
4		
5		
6		
7		

Prepare your slide show presentation. Write a script that you will read for each slide.

For each slide, include:

1. the type of extreme weather that is pictured.

2. a description of the photograph.

3. an interesting fact.

4. what you recall about that type of weather if you have experienced it.

Communicate Information

1. How were other presentations the same as yours? How were they different?

_____

_____

_____

_____

_____

Performance Task
Become a Meteorologist

You will use weather maps and other information to write and present a weather forecast.

Ask a Question What question do you hope to answer from your research?

_____

_____

_____

1. Study the weather maps, data charts, and other information.

2. **Record Data** Create a graph showing the information that you have collected.

Use tools and tell us what the weather will be like tomorrow.

3. **Analyze Data** Circle the information you will use to write your forecast.

Communicate Information

1. Write your weather forecast.

_____

_____

_____

_____

⚙ Crosscutting Concepts
Patterns

2. Describe how patterns helped you make your
   weather forecast.

_____

_____

_____

_____

_____

3. **Construct an Explanation** Were you able to answer
   your question?

_____

_____

_____

❓Essential Question
How does weather change?

Think about the photos of extreme weather you saw at the beginning of the lesson. How did the weather change in each of the weather situations?

_____

_____

_____

_____

_____

⚙️ Science and Engineering Practices

Now that you're done with the lesson, share what you did!

Review the "I can . . ." statement you wrote earlier in the lesson. Explain what you have accomplished in this lesson by completing the "I did . . ." statement.

I did _____

_____

_____

_____

_____

Different Climates

 PAGE KEELEY SCIENCE PROBES

Temperature Drop

Mr. and Mrs. Collins live in Tampa, Florida. They enjoy the warm winter weather. The average temperature in February is a mild 70 degrees F. This year there was a week of very cold temperatures. It dropped to 38 degrees one day, and it almost snowed! Mr. and Mrs. Collins were surprised that it got so cold. This is what they said:

Mr. Collins: *Wow! That is a big change in our weather!*

Mrs. Collins: *Wow! That is a big change in our climate!*

Which person do you agree with more? _____

Explain why you agree.

_____

_____

_____

Science in Our World

Look at the photos of two different climates.
What questions do you have?

_____

_____

_____

_____

Read about a hydrologist and answer
the questions on the next page.

> A hydrologist
> analyzes how precipitation
> changes might affect
> people and places.

STEM Career Connection
Hydrologist

I am a special type of scientist called
a hydrologist. A meteorologist studies water in
the atmosphere. I study Earth's water cycle when
it lands on Earth until it goes back to the atmosphere.

I spend most of my days trying to solve
water-related problems. I think about problems
related to the amount of water in an area, especially
if there is too much. I work with engineers to analyze
the situation and create solutions. Often we create
dams or reservoirs to control or prevent flooding issues.

When a meteorologist is forecasting a lot of rain,
I interpret and analyze the data to see if I need to
solve any problems because of it!

HUGO
Meteorologist

1. Which science and engineering practice does the hydrologist do?

_____

_____

2. How do hydologists work with others to solve problems?

_____

_____

_____

? Essential Question

How do climates vary in different regions of the world?

_____

_____

_____

⚙ Science and Engineering Practices

I will analyze and interpret data.

I will obtain, evaluate, and communicate information

A hydrologist analyzes and interprets data about the water cycle.

Inquiry Activity
Compare Weather Patterns

How do weather patterns in different cities compare?

Ask a Question What question do you hope to answer?

_____

_____

_____

_____

Materials

☐ map or globe

☐ paper

☐ markers

1. Locate the cities of Sitka and Phoenix on a map or globe. Research their average yearly temperatures and precipitation, and other weather data, such as types of precipitation and wind speed.

2. **Record Data** Fill in the table with your findings.

City	Average Yearly Temperatures High/Low	Yearly Precipitation	Other Weather Data
Sitka, Alaska			
Phoenix, Arizona			

3 Analyze Data Circle the highest temperature. Underline the lowest amount of rainfall precipitation. Did either city have other types of precipitation?

_____

_____

_____

Communicate Information

1. How do the temperatures and precipitation amounts compare between the two cities?

_____

_____

_____

2. **Draw Conclusions** Were you able to answer the question you asked?

_____

_____

_____

Obtain and Communicate Information

abc Vocabulary

Use these words when explaining different climates.

climate   axis   season

Climate

📖 Read pages 164–165 in the *Science Handbook.* Answer the following questions after you have finished reading.

1. What is climate?

_____

_____

_____

2. Are climates the same in all locations?

_____

_____

3. Why does point A on the map on page 165 have a warmer climate than point B?

_____

_____

_____

Climate Tour

▶ Watch *Climate Tour* on different types of climates.
Answer the questions after you have finished watching.

4. Give an example of a climate and how it affects what
people do.

_____

_____

5. What is the climate like where you live?

_____

_____

Factors That Affect Climate

📖 Read pages 166–167 in the *Science Handbook*. Answer
the following questions after you have finished reading.

6. Why do many places near the ocean have milder
climates than those farther away from the beach?

_____

_____

7. How do mountains affect a climate?

_____

_____

FOLDABLES®

Cut out the Notebook Foldables tabs given to you by your teacher. Glue the anchor tabs as shown below. Compare the two climates shown in the pictures.

Glue anchor tab here

Comparing Data

Investigate finding patterns in climate data by conducting the simulation. Complete the data chart after you have finished watching.

8. Choose two of the cities that are close to the 30° South latitude line.

   Complete the data chart.

City	Average High Temperature	Average Low Temperature	Warmest Month

Crosscutting Concepts
Patterns

9. What pattern do you see in the data about the two cities?

_____

_____

_____

10. Choose two different cities. Complete the data chart.

City	Average High Temperature	Average Low Temperature	Warmest Month

Crosscutting Concepts
Patterns

11. What pattern do you see in the data about these two cities?

_____

_____

_____

12. Compare your data charts with a classmate's chart. What patterns do you see?

_____

_____

_____

_____

Seasons

📖 Read pages 162–163 in the *Science Handbook.* Answer the following questions after you have finished reading.

13. Summarize what seasons are.

_____

_____

_____

_____

⚙️ Science and Engineering Practices

Think about how you analyzed and interpreted data and obtained information about different climates. Tell how you can analyze, interpret, and communicate by completing the "I can . . ." statements below.

Use examples from the lesson to explain what you can do!

I can _____

_____

I can _____

_____

_____

Research, Investigate, and Communicate

 ## Inquiry Activity
Land and Temperature Change

You will explore what type of land changes temperature the most over a period of time.

Materials
- [ ] soil
- [ ] rocks
- [ ] sand
- [ ] 2 ther-mometers

Write a Hypothesis **If** _____

_____ then _____

_____

_____ because _____

_____

Carry Out an Investigation

① Plan your procedure.

_____

_____

_____

_____

_____

2 **Record Data** Create a table to show the data
you collect.

3 **Analyze Data** Circle the type of land that warmed up
the most in the given time.

Communicate Information

1. How could the type of land in an environment affect
the local climate?

_____

_____

_____

2. **Construct an Explanation** Were you able to prove
your hypothesis?

_____

_____

Performance Task
Create a Climate Travel Poster

You will create a travel poster that features climate information.

Materials
☐ colored pencils
☐ books and resources

Ask a Question What question would you expect to answer from your research?

_____

_____

_____

1. Choose Sitka, Alaska, or Phoenix, Arizona.

2. Review the information that you have gathered in this lesson or collect information from other sources.

3. Include what the climate of your city is like including various temperatures or features of that climate. You may also want to include the activities that are possible in that city because of its climate.

4. Create a draft of your travel poster. Record the information that you will include.

A hydrologist considers climate when thinking about water problems and solutions. You might think about climate when you want to go on a trip!

_____

_____

_____

_____

_____

Communicate Information

1. Draw your travel poster.

? Essential Question

How do climates vary in different regions of the world?

Think about the photos of Sitka, Alaska, and Phoenix, Arizona, at the beginning of the lesson. How are the climates different in these two regions?

_____

_____

_____

⚙ Science and Engineering Practices

Now that you're done with the lesson, share what you did!

Review the "I can . . ." statements you wrote earlier in the lesson. Explain what you have accomplished in this lesson by completing the "I did . . ." statements.

I did _____

_____

I did _____

_____

_____

Weather and Climate

Performance Project
Five-Day Forecast

Look back at the questions that you wrote at the beginning of the module. Now that you have learned about weather and climate, you can probably answer many of them.

Meteorologists study the weather and climate. Assume the role of a meteorologist. Research the difference between the weather and the climate of your favorite city. Record the weather and climate information that you find below or organize it in a chart.

How does a meteorologist forecast the weather as far as five days in advance?

Based on this information, what do you think the weather might be like one week from now in your favorite city? What might it look like in one month? Explain your thinking. Communicate your findings to your class.

_____

_____

_____

_____

_____

_____

_____

_____

_____

 ## Explore More in Our World

Did you learn the answers to all of your questions from the beginning of the module? If not, how could you design an experiment or conduct research to help answer them?

Parents and Offspring

 ## Science in Our World

Look at the photo. Cicadas are strange looking insects!
Share what you have observed about them. What questions
do you have about cicadas after observing the photo?

_____

_____

_____

abc Key Vocabulary

Look and listen for these words as you learn
about parents and offspring.

birth	germinate	inherit
instinct	life cycle	metamorphosis
offspring	reproduction	trait
variation		

How is the life cycle of a cicada different from other insects?

OWEN
Entomologist

STEM Career Connection
Entomologist

When you are an entomologist, every day is different! We have many important tasks, such as studying insect life cycles, behaviors, and populations. We also classify insects into groups to learn more about them and how they are similar to each other. We work at places like zoos, parks, and science labs. Some entomologists might work in a zoo on some days and in a science lab on other days.

What do you think the life cycle of a cicada is?

Science and Engineering Practices

I will develop and use models.

I will construct explanations.

I will analyze and interpret data.

Life Cycle of Plants

PAGE KEELEY SCIENCE PROBES

Life Cycles

Group A	Group B	Group C	Group D	Group E
frogs	people	butterflies	crabs	trees
chickens	elephants	moths	worms	grass
fish	tigers	beetles	snails	bean plants

Plants and animals are grouped in the table shown above.
How many of the groups contain examples of things
that have a life cycle? Circle the answer that best matches
your thinking.

A. None of the groups

B. 1 group

C. 2 groups

D. 3 groups

E. 4 groups

F. All 5 of the groups

Explain your thinking. How did you decide if something had
a life cycle?

_____

_____

_____

_____

Science in Our World

▶ Watch the video of the plants growing. What questions do you have?

_____

_____

_____

_____

Read about a botanist, and answer the questions on the next page.

A botanist needs to understand the life cycles of the plants he studies.

STEM Career Connection
Botanist

Some people call me a plant scientist. It's true, I study plants, their needs, and their life cycles. I've been studying a type of tree called the live oak. One tree that I studied was over 500 years old. It was a very important tree to many people in the community.

That tree got sick and began to die. We tried many different things to save it. We dug up the ground around the roots and put in nutrient-rich soil. We also used water sprinklers on timers so the tree got the water that it needed each day. The tree began to recover and soon was making acorns again. Acorns are its seeds. It was the best experience I've had as a botanist.

POPPY
Park Ranger

1. What are the seeds of a live oak tree called?

_____

2. What helped the old tree get healthy?

_____

_____

? Essential Question
How are life cycles of plants similar?

_____

_____

_____

⚙ Science and Engineering Practices

I will develop and use models.

Like a scientist, you will develop and use models to show what you know.

Inquiry Activity
Seed Growth

What do seeds need in order to grow?

Make a Prediction Which seed do you think will grow, one that has water or the one that does not?

_____

_____

_____

Materials

☐ 2 sandwich bags

☐ 2 paper towels

☐ bean seeds

☐ water

Carry Out an Investigation

1. Fold each paper towel so that it fits inside a sandwich bag.

2. Soak one of the paper towels with water. Place the wet paper towel in one bag, and lay it flat. Place the dry paper towel in the other bag.

3. Place some seeds on the paper towel in each bag. Press each seed firmly into the paper towel.

4. Seal each of the bags, and place them next to each other in a safe place.

5. **Record Data** Observe the seeds daily. Record your observations in the table on the next page.

Seed Observations	
Day 1	
Day 2	
Day 3	
Day 4	
Day 5	

Communicate Information

1. What differences did you notice between the two bags of seeds?

_____

_____

_____

2. Based on your observations, what do you think seeds need in order to grow?

_____

_____

Obtain and Communicate Information

🔤 Vocabulary

Use these words when explaining plant life cycles.

germinate   **life cycle**   **reproduction**

Plant Needs

📖 Read pages 54–55 in the *Science Handbook*.
Answer the following questions after you have
finished reading.

1. What are five things that plants need in order to live?

_____

_____

2. What role do plants play in creating oxygen that animals
   need to breathe?

_____

_____

_____

3. Why do you think big trees in a forest grow far apart
   from each other?

_____

_____

Inquiry Activity
Parts of a Seed

You observed some of the things seeds need to grow. Now examine the parts of a seed.

Ask a Question Write a question that you would like to answer in your investigation.

Materials

☐ water

☐ bean seed

☐ magnifying glass

Glue observations here.

_____

_____

Carry Out an Investigation

① Soak the bean seed in water for 24 hours before the investigation.

② Pick up the bean. Examine it using the magnifying glass. On a separate sheet of paper, draw what you see.

③ Rub the soaked bean between your fingers. The outer covering should rub off. What does the bean look like now? Write a description below your drawing.

④ Locate the line that runs down the middle of the seed. Split the seed apart along that line.

⑤ Use the magnifying glass to examine the inside of the seed. Draw what you see on the sheet of paper you used in step 2.

From Seed to Plant

📖 Read page 58 in the *Science Handbook*. Answer
   the following questions after you have finished reading.

4. Use the information you've read to label the drawings
   you made during the investigation.

5. What is the vocabulary word used to describe when
   seeds start sprouting?

_____

Reproducing with Flowers

📖 Read page 56 in the *Science Handbook*. Answer
   the following questions after you have finished reading.

6. Describe the process of pollination.

_____

_____

_____

_____

7. What is the role that fruit plays in a plant life cycle?

_____

_____

⚙️ Crosscutting Concepts
Patterns

8. What is the next step in the life cycle of a cherry tree
   after it has grown into an adult?

_____

_____

Seed Growth

Revisit the *Seed Growth* activity on pages 129–130 in
the ***Be a Scientist Notebook***. Use what you have learned
in the lesson so far to answer the following questions.

9. Which part of the plant life cycle are you seeing in the
   Seed Growth activity?

_____

_____

_____

10. What would happen next in the life cycle of this plant?

_____

_____

11. Draw a diagram to show what happened to the seed as
    it started to grow.

FOLDABLES

Cut out the Notebook Foldables tabs given to you by
your teacher. Glue the anchor tabs as shown below.
Use what you have learned to define each word using
the picture of the pumpkin life cycle.

Glue anchor tab here

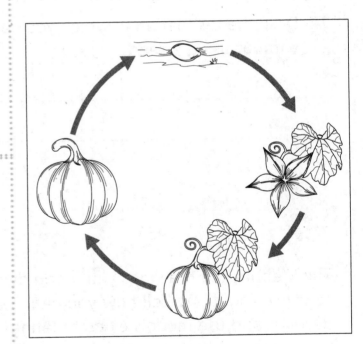

Cones and Spores

📖 Read pages 57 and 59 in the *Science Handbook.* Answer the following questions after you have finished reading.

12. Where do the seeds of pine trees form?

_____

_____

13. What is the difference between a seed and a spore?

_____

_____

14. Describe a situation in which you can grow a new plant without using a seed.

_____

_____

Science and Engineering Practices

Think about the drawings you made during your investigations. Tell how you can develop and use models by completing the "I can . . ." statement below.

Use examples from the lesson to explain what you can do!

I can _____

_____

_____

Research, Investigate, and Communicate
Seed Dispersal

Explore the Digital Interactive *Seed Dispersal* on how plants spread their seeds. Answer the questions after you have finished.

1. How are dandelion seeds spread?

_____

_____

2. Share two ways animals help with seed dispersal.

_____

_____

_____

3. Think about the needs of plants. Why might seed dispersal be important?

_____

_____

4. What do you think the word *dispersal* means?

_____

_____

Performance Task
Plant Life Cycle Model

A botanist needs to understand the life cycles of plants. You will use what you learned in the lesson to draw a model of a plant's life cycle.

Ask a Question Write a question that you would like to be able to answer with your model.

_____

_____

_____

Make a Model

1. Watch the video from the beginning of the lesson.

2. The video shows part of the life cycle of a flowering plant. Look at the diagram on page 56 in the *Science Handbook.* The diagram shows the complete life cycle of a cherry tree.

3. Using the diagram on page 56 as an example, draw a model of the life cycle of the plant shown in the video.

4. Include and label the following stages in the life cycle in your model: seed, germination, young plant, flower, pollination, and adult plant.

Be like a botanist and create a model of the flowering plant's life cycle.

Communicate Information

1. **Construct an Explanation** How is the life cycle of this plant similar to the life cycle of a plant that reproduces with cones? How is it different?

_____

_____

_____

❓ Essential Question
How are life cycles of plants similar?

▶ Think about the video of the growing plant you watched. Explain how the life cycles of plants are similar.

_____

_____

⚙️ Science and Engineering Practices

Review the "I can . . ." statement you wrote earlier in the lesson. Explain what you have accomplished in this lesson by completing the "I did . . ." statement.

Now that you are done with the lesson, share what you did!

I did _____

_____

_____

Life Cycles of Animals

PAGE KEELEY
SCIENCE
PROBES

Life Cycle Stages

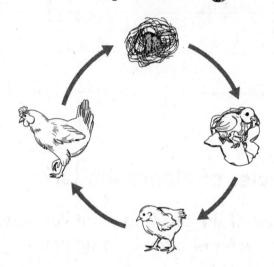

Which animals go through a life cycle similar to the stages of a chicken's life cycle?

Butterfly	Horse	Duck
Snake	Turtle	Lizard
Cat	Crow	Mouse

Explain your thinking. How did you decide if the stages of an animal's life cycle are similar to the stages of a chicken's life cycle?

_____

_____

_____

_____

Science in Our World

Look at the photo of a duck-billed platypus. Think about what it might have looked like as a baby. What questions do you have?

_____

_____

_____

_____

Read about a wildlife biologist and answer the questions on the next page.

> A wildlife biologist needs to understand the life cycles of the animals she studies.

STEM Career Connection
Wildlife Biologist

My job as a wildlife biologist lets me work with many interesting animals. I spent some time in Australia where I studied the duck-billed platypus. The duck-billed platypus is a fascinating mammal. I spent time with them in their natural habitat to learn their behaviors. I learned about their life cycle, which is different from most other mammals. I took several pictures and wrote notes so that I could share what I learned with others. Some days, I work in the office and use other scientists' research to learn more about the duck-billed platypus.

POPPY
Park Ranger

1. In which part of the world did the wildlife biologist study
   the duck-billed platypus?

   _____

   _____

2. What does the wildlife biologist do in the office?

   _____

   _____

   _____

? Essential Question
How are life cycles of animals similar?

_____

_____

_____

Science and Engineering Practices

I will construct explanations.

Like a scientist, you will construct explanations to show what you learned.

✋ Inquiry Activity
Hatching Brine Shrimp

What is the life cycle of brine shrimp? Brine shrimp eggs hatch when they are in the right conditions. Think about animals that you know hatch from eggs.

Make a Prediction How long will it take brine shrimp eggs to hatch?

_____

_____

Carry Out an Investigation

① Place the masking tape halfway up the outside of the cup. Write your name on the tape.

② Pour the salt water into the cup up to the line created by the masking tape.

③ Place a pinch of brine shrimp eggs into the salt water. Use the craft stick to stir the water until the eggs are no longer floating.

④ Place the cups in a safe, warm place under a light. Do not place them near a window.

⑤ Each day, examine the contents of the cup with the hand lens. After the shrimp begin to hatch, add a pinch of yeast powder to the cup twice daily. This provides food for the shrimp.

⑥ Cover the cup with plastic wrap. Secure with a rubber band. Poke two small holes in the plastic wrap.

⑦ Refill the cup to the masking tape line with tap water as needed. Do not use salt water to refill the cup.

Materials

- ☐ clear plastic cup
- ☐ masking tape
- ☐ marker
- ☐ salt water
- ☐ brine shrimp eggs
- ☐ craft stick
- ☐ hand lens
- ☐ yeast
- ☐ plastic wrap
- ☐ rubber band

8 **Draw a picture of what the brine shrimp look like after they have hatched.**

Communicate Information

1. Can you think of any other animals that hatch from eggs? How are those eggs different from brine shrimp eggs?

_____

_____

_____

2. Was your prediction correct? How long did it take the brine shrimp eggs to hatch?

_____

_____

💬 Obtain and Communicate Information

🔤 Vocabulary

Use these words when explaining animal life cycles.

offspring   metamorphosis   birth

Life Cycles

▶ Watch *Life Cycles* on plant and animal life cycles.
Answer the question after you have finished watching.

1. What do life cycles of plants and animals have in common?

_____

_____

Animal Reproduction

📖 Read page 75 in the **Science Handbook**. Answer
the following questions after you have finished reading.

2. What is one example of reproduction from one parent?
Describe what happens in your example.

_____

_____

_____

3. What is another word for the babies that parent animals
create during reproduction?

_____

Animal Life Cycles

📖 Read pages 71–72 in the *Science Handbook*. Answer
the following questions after you have finished reading.

4. Which two animal groups undergo metamorphosis as
part of their life cycles?

_____

5. What do the life cycles of amphibians and insects have
in common?

_____

_____

_____

Butterfly and Salamander Life Cycles

🔊 Explore the Digital Interactive *Butterfly and Salamander
Life Cycles* on the life cycles of these insects and
amphibians. Answer the following questions after you
have finished.

6. At what stage in its life cycle does a caterpillar change
into a butterfly?

_____

_____

7. How does the larva of a salamander change as it grows?

_____

_____

Life Cycles of Reptiles, Fish, and Birds

📖 Read page 73 in the *Science Handbook*. Answer
the following questions after you have finished reading.

8. How do the life cycles of most birds, fish, and reptiles
begin? Which other animal groups also begin life this
way?

_____

_____

_____

9. What are some differences in life cycles of birds, reptiles,
and fish?

_____

_____

_____

Salmon Life Cycle

Explore the Digital Interactive *Salmon Life Cycle* on
the life cycle of this type of fish. Answer the following
questions after you have finished.

10. How are the habitats of adult salmon and young
salmon different?

_____

_____

_____

11. What does the word *spawn* mean?

_____

Life Cycles—Mammals

📖 Read page 74 in the *Science Handbook*. Answer
the following questions after you have finished reading.

12. What is the main difference between mammal life
cycles and the life cycles of birds, fish, and reptiles?

_____

_____

13. How do mammal parents care for their young?

_____

_____

14. How do you think young chimpanzees learn how to
climb trees? Why do you think so?

_____

_____

_____

_____

⚙️ Crosscutting Concepts
Patterns

15. A pattern is something that repeats. Do you think
a life cycle is a pattern? Explain.

_____

_____

_____

_____

FOLDABLES®

Cut out the Notebook Foldables tabs given to you
by your teacher. Glue the anchor tabs as shown below.
Use what you have learned about life cycles to
describe the life cycle of the lion.

Glue anchor tab here

Glue anchor tab here

Elephant and Horse Life Cycles

Explore the Digital Interactive *Elephant and Horse Life Cycles* on the life cycle of these mammals. Answer the following questions after you have finished.

16. What are the names of baby elephants and baby horses?

_____

_____

17. How are the life cycles of elephants and horses similar?

_____

_____

_____

18. Which mammal has a longer life cycle?

_____

_____

Science and Engineering Practices

Use examples from the lesson to explain what you can do!

Think about the ways you have explained your answers in this lesson. Tell how you can construct explanations by completing the "I can . . ." statement below.

I can _____

_____

Research, Investigate, and Communicate

Life Cycle Rule Breakers

Read the Science File *Life Cycle Rule Breakers* on animals that do not follow the normal life cycle pattern for their animal group. Answer the questions after you have finished reading.

1. How is the life cycle of a garter snake different than most reptiles?

_____

_____

2. Which animal in the Science File has the largest number of offspring?

_____

_____

3. A koala is a mammal that has a pouch to carry its young. Do you think a koala is a marsupial?

_____

_____

4. How is the life cycle of a seahorse similar to the life cycles of most fish?

_____

_____

Performance Task
Duck-Billed Platypus Life Cycle Model

Materials

☐ *Science Handbook*

☐ research materials

A wildlife biologist needs to understand the life cycles of the animals that he or she studies. The duck-billed platypus is another mammal that is a life cycle rule breaker. You will research the life cycle of a duck-billed platypus and create a model of its life cycle.

Ask a Question Write a question that you would like to be able to answer with your model.

_____

_____

_____

_____

Make a Model

1 Reread page 74 in the *Science Handbook* on the life cycles of mammals.

2 Research the life cycle of a duck-billed platypus. Record your notes below.

_____

_____

_____

_____

_____

③ Using the diagram on page 74 of the *Science Handbook* as an example, draw a model of the life cycle of the duck-billed platypus.

④ Label the important stages of the life cycle.

Communicate Information

1. **Construct an Explanation** How is the life cycle of the duck-billed platypus different than the life cycles of most other mammals?

_____

_____

_____

2. **Construct an Explanation** How is the life cycle of the duck-billed platypus like the life cycles of most other mammals?

_____

_____

_____

_____

3. What questions do you have after researching duck-billed platypuses? Did you learn any interesting facts?

_____

_____

_____

_____

? Essential Question
How are life cycles of animals similar?

Think about the photo of the duck-billed platypus at the beginning of the lesson. Use the duck-billed platypus as an example to explain how life cycles of animals are similar.

_____

_____

_____

⚙ Science and Engineering Practices

Now that you're done with the lesson, share what you did!

Review the "I can . . ." statement you wrote earlier in the lesson. Explain what you have accomplished in this lesson by completing the "I did . . ." statement.

I did _____

_____

_____

_____

Inherited and Learned Traits

Sadie's Poodle

Sadie has a large, white poodle. Her poodle does tricks, barks at strangers, and wags her tail when she is happy. Sadie and her friends have different ideas about the poodle's size, her fur color, and her behaviors. Here is what they said:

Sadie: My poodle gets all of her characteristics from her parents.

Josh: I think your poodle gets some of her characteristics from her parents and some from interacting with her environment.

Omar: I think your poodle gets all of her characteristics from interacting with her environment.

Who do you agree with the most? _____

Explain why you agree.

_____

_____

_____

Science in Our World

Look at the photo of the mice. How are they alike?
How are they different? What questions do you have?

_____

_____

_____

_____

Read about a geneticist and answer
the questions on the next page.

> A geneticist understands how traits get passed down from parents to offspring.

STEM Career Connection
Geneticist

    I am a geneticist, and my job is to study
genetics. I examine how adults pass on traits to their
young through genes. I use different kinds of instruments
in a lab to do most of my work. I gather information
about traits of people, plants, and animals. It is very
important for me to collect and analyze data to continue
learning about different traits. Pictures can reveal many
of the traits that parents pass down to their offspring. If
you look at a picture of a cat and her kittens, you
will probably see that they look alike in many ways, but
not all ways. Parents also pass on traits that affect
the way offspring behave. That's another
fascinating part of genetics that I get to explore in my
work.

NOAH
Nurse

1. What do you think the term *genetics* means?

_____

_____

_____

2. Where does this geneticist do most of his or her work?

_____

_____

? Essential Question
What affects an organism's traits?

_____

_____

Science and Engineering Practices

I will analyze and interpret data.

Like a geneticist, you will analyze and interpret the data you gather.

Inquiry Activity
Graphing Inherited Traits

Which observable traits are the most common? Some observable traits are inherited from parents. You will collect data for three of those traits—freckles, eye color, and hairline— to see which one is the most common.

Make a Prediction Which trait do you think will be the most common in your class? Why?

_____

_____

_____

Carry Out an Investigation

1. Look in a mirror. Do you have freckles on your face? Record whether or not you have freckles.

2. What color of eyes do you have? Record it.

3. Pull your hair back so you can see your forehead. Observe your hairline in a mirror. Is it straight across or does it come to a point? Record what your hairline is like.

Freckles: have or don't have?	
Eye color: brown, blue, or other?	
Hairline: straight or pointed?	

4 **Record Data** Collect data from the entire class. Find how many people have each trait and record it in the table.

	Freckles	No Freckles	Brown Eyes	Blue Eyes	Other Color Eyes	Straight Hairline	Pointed Hairline
Number of Students							

5 On a separate sheet of paper, draw a bar graph to represent the data.

6 **Analyze Data** Use the bar graph to determine which inherited trait was the most common. Write the most common trait below.

_____

_____

Communicate Information

1. Was your prediction about the most common trait correct?

_____

_____

2. Have you noticed traits shared by people who are related? Explain.

_____

_____

Glue your graph here.

Obtain and Communicate Information

Vocabulary

Use these words when explaining inherited and learned traits.

inherit   trait   instinct

variation

Inherited Traits

▶ Watch *Inherited Traits* on genetics. Answer the questions after you have finished watching.

1. What does the word *inherit* mean?

_____

_____

2. What is an example of an inherited trait that you cannot see?

_____

_____

Inherited Physical Traits and Instincts

📖 Read page 76 in the *Science Handbook*. Answer the following question after you have finished reading.

3. Where do offspring get their inherited traits?

_____

_____

Inherited Physical and Behavioral Traits

👁 Read the Science File *Inherited Physical and Behavioral Traits* to learn about the traits of one type of mammal. Answer the question after you have finished reading.

4. Describe a way in which the physical traits and the behavioral traits of the bushbaby are related.

_____

_____

_____

Variety of Inherited Traits

🔄 Explore the Digital Interactive *Variety of Inherited Traits* to learn more about inherited traits in animals and plants. Answer the following questions after you have finished.

5. Describe two inherited traits of a fish. Classify the traits as either a physical trait or an instinct.

_____

_____

⚙ Crosscutting Concepts
Cause and Effect

6. Do the tulips always have the exact same color as their parents? What might cause this? Use evidence from the photo to support your thoughts.

_____

_____

_____

FOLDABLES

Cut out the Notebook Foldables tabs given to you by
your teacher. Glue the anchor tabs as shown below.
Use what you learned about inherited traits for each
big cat to determine how inherited traits affect its fur.

Glue anchor tab here

Learned and Environmental Traits

📖 Read page 77 in the *Science Handbook*. Answer the following question after you have finished reading.

7. How is a learned trait different from an inherited trait?

_____

_____

_____

▦ Investigate how the environment affects a plant by conducting the simulation. Answer the question after you have finished.

8. How did the environment affect the plant?

_____

_____

_____

⚙ Science and Engineering Practices

Use examples from the lesson to explain what you can do!

Think about the ways you have analyzed and interpreted data in this lesson. Tell how you can analyze and interpret data by completing the "I can . . ." statement below.

I can _____

_____

🔍 Research, Investigate, and Communicate
Pea Plants

👁 Read the Science File *Pea Plants* on the history of genetics. Answer the questions after you have finished reading.

1. What observation did Gregor Mendel make that led to the science of genetics?

_____

_____

_____

2. What trait did Mendel study in pea plants?

_____

3. What happened when plants with purple flowers reproduced with plants with white flowers?

_____

_____

4. What does the word *dominant* mean?

_____

_____

⚙ Performance Task
Mouse Fur Color Inheritance

Think about the two mice at the beginning of the lesson.
Each had a different fur color. Conduct research to find out
whether fur color in mice is an inherited trait.

Ask a Question What question will your research help to
answer?

_____

_____

_____

Communicate Information

1 **Make an Argument** Is fur color an inherited trait
   in mice? Use the information you found in your research
   as evidence to support your argument.

_____

_____

_____

_____

2 Do you think that the mice in the picture
   have any other physical inherited traits?
   What are they?

_____

_____

_____

_____

❓ Essential Question
What affects an organism's traits?

Think about the photo of the gray mouse and the white mouse. Use the mice as an example to explain the things that can affect an organism's traits.

_____

_____

_____

_____

_____

⚙️ Science and Engineering Practices

Now that you're done with the lesson, share what you did!

Review the "I can . . ." statement you wrote earlier in the lesson. Explain what you have accomplished in this lesson by completing the "I did . . ." statement.

I did _____

_____

_____

Parents and Offspring

⚙ Performance Project
Comparing Life Cycles

Scientists are always discovering new things about life cycles. Because of entomologists, we are able to see fascinating pictures and learn amazing things about all different types of insects. Use print and digital resources to research the life cycle of a cicada. They really are fascinating creatures! Take notes on a separate sheet of paper.

In the boxes below, develop a model to show the life cycle of a cicada in one box and the life cycle of another animal from this module in the other box. Label your models.

Cicada Life Cycle	_____ Life Cycle

Use what you've learned about life cycles to find out about cicadas.

Scientists often use diagrams, illustrations, or 3D models to make comparisions. How are the life cycles of the two animals the same? How are the life cycles of the two animals different? What are some ways to further study the life cycles of various animals?

_____

_____

_____

_____

_____

_____

 # Explore More in Our World

Did you learn the answers to all of your questions from the beginning of the module? If not, how could you design an experiment or conduct research to help answer them?

Survival

Science in Our World

Honeybees live in groups called colonies. A colony lives in a hive. The bees that you see collect food for the colony. Each bee in the colony has an important job. Look at the photo of honeybees. What questions do you have?

_____

_____

_____

_____

🔤 Key Vocabulary

Look and listen for these words as you learn about survival.

adaptation	camouflage	competition
hibernation	migration	mimicry
natural selection	population	trait

How do bees benefit
from working in groups?

OWEN
Entomologist

STEM Career Connection
Beekeeper

Some beekeepers gather honey as a hobby. I collect honey from
my bees and sell it. It is important that I understand how bees
work and live together. I have spent time researching and learning
what I can about bees. During the spring and summer months,
the bees collect nectar to produce food for the colony. When it is
time to collect the honey, I remove the honeycomb and honey from
each hive. I wear protective clothing so I am not stung. Sometimes
I use smoke to make the bees calm as I gather the honey.

What questions do you have for the beekeeper?

_____

_____

_____

_____

⚙️ Science and Engineering Practices

I will analyze and interpret data.

I will engage in argument from evidence.

I will construct explanations.

Animal Group Survival

PAGE KEELEY
SCIENCE
PROBES

Animal Groups

Many animals interact with each other in groups. Put an X in all the boxes that describe how animals interact in groups.

Animals interact in groups when they need to find food.	Animals interact in groups when they need to defend themselves.
Animals interact in groups when they need to cope with changes.	Animals sometimes interact in small groups.
Animals sometimes interact in large groups.	Animals only interact in pairs.
Animals only interact with animals of their own kind.	Animals sometimes interact with other kinds of animals to survive.

Explain your thinking. Describe your ideas about how animals interact in groups.

_____

_____

_____

_____

Science in Our World

📄 Explore the Digital Interactive *Animal Groups*. Why might animals live in groups? What questions do you have?

_____

_____

_____

_____

Read about a marine biologist and answer the questions on the next page.

> A marine biologist studies many animals that live in groups in the oceans.

POPPY
Park Ranger

○ **STEM Career Connection**
○ ## Marine Biologist
○   During the month of June, I studied the behavior
○ of Pacific sardines. I went diving to observe several
○ schools of sardines. As predators approached, the
○ school swam faster and swirled through the water.
○ It looked like whirling smoke or a tornado. Sometimes
○ they formed a huge circle. The sardines did this
○ for protection and to confuse predators.

1. What do you think a group of fish is called?

_____

2. Why did the group of sardines move the way they did?

_____

_____

_____

? Essential Question

How does being part of a group help animals survive?

_____

_____

_____

Science and Engineering Practices

I will construct explanations.

Like a scientist, you will construct explanations to tell why something happens.

Inquiry Activity
Ant Workers

Materials

☐ craft sticks

☐ timer

How does working in a group help an ant colony?
Ants work together to build bridges and trails. You
will build on your own and in a group and
compare the results.

Make a Prediction You will have 1 minute to build
a road out of craft sticks across a desk. Will you be
able to build a better road by yourself or as a
team? Explain.

_____

_____

_____

_____

Carry Out an Investigation

1. Place the craft sticks in a pile on one side of the room.
   Stand on the other side of the room.

2. Set the timer for 1 minute.

3. Start the timer. Walk to the pile of craft sticks and bring
   one back at a time. Place it on your desk to begin
   building a road across it. Repeat as many times as
   possible until time is up.

4. **Record Data** Count the craft sticks and describe your
   road. Record the data in the table on the next page.

5. Repeat steps 2–4 but have four students gather craft
   sticks and work on the road at the same time.

6. Repeat step 5, but this time try to work together in a
   different way.

Number of Students	Number of Craft Sticks in Road	Description of Road
1		
4		
4		

Communicate Information

1. Was your prediction correct? How did the road that you built by yourself compare to the road that you built as a team?

_____

_____

_____

2. What benefits are there to working as a team in an ant colony? Use your data to construct an explanation.

_____

_____

_____

3. What different way did you try to work as a team?

_____

_____

Obtain and Communicate Information

abc Vocabulary

> Use these words when explaining animal groups.
>
> group   survive   population

Animal Groups

▶ Watch *Animal Groups* on the way different animals work together in groups. Answer the questions after you have finished watching.

1. What are some advantages of living in a group?

_____

_____

_____

2. Choose two animals from the video. How are these two groups the same? How are they different?

_____

_____

_____

3. Choose an animal group from the video. Think about how your family members take care of each other. How is your family like the animal group? How is it different?

_____

_____

_____

Animal Groups

👁 Read *Animal Groups* on animals that work together and animals that work alone. Answer the following question after you have finished.

4. For some animals it is better to live in groups. What are the advantages for those animals who live in groups?

_____

_____

_____

⚙ Crosscutting Concepts
Cause and Effect

5. What do you think would happen if an animal became separated from its group?

_____

_____

_____

An Elephant Herd

📖 Explore the Digital Interactive *An Elephant Herd* on how elephants work together. Answer the question after you have finished.

6. Describe the roles within the elephant herd.

_____

_____

_____

_____

_____

Who Works Together? Who Works Alone?

👁 Read *Who Works Together? Who Works Alone?* on how some animals live in groups and others live alone. Answer the following questions after you have finished.

7. Describe the differences between an animal that lives in a group and one that lives alone.

_____

_____

_____

_____

_____

_____

8. What are the advantages of hunting alone instead of hunting as a group?

_____

_____

9. Explain why certain animals live together and others live alone.

_____

_____

_____

_____

_____

Name _____ Date _____

FOLDABLES

Cut out the Notebook Foldables tabs given to you by your teacher. Glue the anchor tabs as shown below. Use what you have learned to describe and compare these two animal groups.

Glue anchor tab here

Glue anchor tab here

All Different Sizes

👁 Read *All Different Sizes* on numbers of animals in animal groups. Answer the following questions after you have finished.

10. Describe the two needs that affect the size of a wolf pack.

_____

_____

_____

_____

11. Do you think zebras or lions live in larger groups? Explain.

_____

_____

_____

_____

⚙️ ## Science and Engineering Practices

Use examples from the lesson to explain what you can do!

Think about the explanations you wrote. Tell how you can construct explanations by completing the "I can . . ." statement below.

I can _____

_____

_____

Research, Investigate, and Communicate
Minnow Observations

You have learned about the ways animals benefit from living in groups. You can see group behavior in action by studying minnows.

Make a Prediction What do you think will happen if you introduce a new minnow into a school of minnows? Why do you think this?

_____

_____

_____

Materials
☐ clear container
☐ aerator
☐ water
☐ cup
☐ minnows
☐ fish food

Carry Out an Investigation

BE CAREFUL Always handle live animals with care. Wash your hands after handling.

1. Fill a clear container with water. Let the water sit out for one day before adding the minnows. Set up the aerator in the container.

2. Scoop some of the water from the container into the cup.

3. Gently place one minnow in the cup. Place the rest of the minnows in the container of water.

4. Observe the behavior of the school of minnows.

5. Slowly pour the minnow from the cup into the container with the rest of the minnows.

6. Observe the behavior of the new minnow.

7. Add some fish food to the container. Observe the behavior of the minnows when they receive food.

8. Place your hand in the water near the minnows. Observe the way the minnows react to your hand.

Communicate Information

1. Describe the behavior of the school of minnows.

_____

_____

_____

2. What happened when you added the new minnow to the container? Did the results support your hypothesis?

_____

_____

_____

_____

3. Describe the minnows' behavior when food was added to the container.

_____

_____

4. What happened when you placed your hand in the container? What might cause the minnows to react this way in the wild?

_____

_____

_____

Performance Task
Animal Group Explanation

Recall the animal groups that you have studied
in this lesson or use the Digital Interactive
Animal Groups from Engage. Construct an explanation
of how living in a group helps one of those animals
survive. Use observations and data as evidence
to support your explanation.

Communicate Information

_____

_____

_____

_____

_____

_____

_____

_____

_____

_____

❓ Essential Question
How does being part of a group help animals survive?

📖 Think about the photos of animal groups at the beginning of the lesson. Use those examples to explain how living in a group helps some animals survive.

_____

_____

_____

_____

⚙️ Science and Engineering Practices

Now that you are done with the lesson, share what you did!

Review the "I can . . ." statement you wrote earlier in the lesson. Explain what you have accomplished in this lesson by completing the "I did . . ." statement.

I did _____

_____

_____

Adaptations

PAGE KEELEY
SCIENCE
PROBES

Adaptations

Polar bears live in the cold Arctic. They grow a coat of thick fur to stay warm.

Two friends were at a zoo in Florida. The zoo had a polar bear exhibit. They wondered how the polar bear could live in Florida where it is very warm. This is what they said:

Suzanne: *The polar bear will try to adapt by growing less fur.*

Milo: *The polar bear will not try to adapt by growing less fur.*

Who do you agree with the most? _____

Explain why you agree.

_____

_____

_____

_____

Science in Our World

Explore the Digital Interactive *Different Birds* and observe the photos. The birds are similar to each other in some ways, and very different in others. What questions do you have?

_____

_____

_____

_____

Read about an ornithologist and answer the questions on the next page.

An ornithologist studies bird adaptations.

STEM Career Connection
Ornithologist

I am an ornithologist—a scientist who studies birds. There are thousands of different types of birds. I get to travel around the world to study the traits of birds. Birds are different in different places. My favorite place is the rain forest. In South America, I studied the tiny bee hummingbird. It is only about 6 centimeters long. In a rain forest on the other side of the world, I studied the Australian cassowary. Those birds can be up to 2 meters tall. That's taller than many adult humans! Birds also have many different colors, beaks, and wing sizes. Each bird has traits that help it survive in its environment. For instance, their beak style will help them find or eat food.

HIRO
Ocean Engineer

1. What is one of the differences between a bee hummingbird and a cassowary?

_____

_____

_____

2. What does the ornithologist say that all birds have in common?

_____

_____

_____

? Essential Question

How do adaptations help plants and animals survive?

_____

_____

_____

 ## Science and Engineering Practices

I will engage in argument from evidence.

Like a scientist, you will use evidence to support your ideas.

Inquiry Activity
Bird Beak Adaptations

How does beak shape affect a bird? You will observe how birds with different types of beaks gather food by creating a model of the beaks. Read the investigation before you make a prediction.

Make a Prediction How does the shape of a bird's beak affect its ability to gather different types of food?

_____

_____

_____

Materials
☐ plastic spoon
☐ tweezers
☐ large binder clip
☐ scissors
☐ cardboard box lids or trays
☐ paper clips
☐ rubber bands
☐ toothpicks
☐ dried macaroni
☐ plastic cups

Carry Out an Investigation

1. Collect a plastic spoon, tweezers, a binder clip, and a pair of scissors. These will represent the bird beaks.

2. Place four cardboard box lids in front of you. Fill each lid with one of the following: paper clips, rubber bands, toothpicks, and dried macaroni. These will represent bird food.

3. Hold a plastic cup in one of your hands and one of the bird beak models in the other hand. The plastic cup will represent a bird's stomach.

4. Your teacher will set a timer for 20 seconds. When the timer starts, use one bird beak model to collect as much of one type of food as possible. Collect food by using the beak to move the food into the cup.

5. **Record Data** Count the amount of food you were able to collect. Record your data in the table on the next page.

6. Repeat steps 4 and 5 until you have collected each food type with each beak model.

Model Beak	Amount of Food Collected			
	Paper Clips	Rubber Bands	Toothpicks	Macaroni
Spoon				
Tweezers				
Binder Clip				
Scissors				

Communicate Information

1. **Analyze Data** Which bird beak worked best for each type of food? Use evidence to support your ideas.

_____

_____

_____

_____

_____

2. Compare your results with a classmate's results. Are they the same? What do you think it might mean if the results were different?

_____

_____

_____

_____

Obtain and Communicate Information

🔤 Vocabulary

Use these words when explaining adaptations.

adaptation	camouflage	mimicry
hibernation	migration	

Animals in Their Own Environment

▶ Watch *Animals in Their Own Environment* on the things that help animals in different types of environments survive. Answer the question after you have finished watching.

1. What characteristics of the rattlesnake help it survive in the desert?

_____

_____

Adaptations

📖 Read pages 94–95 in the *Science Handbook.* Answer the following questions after you have finished reading.

2. What are the three main ways that adaptations help organisms survive in their environment?

_____

_____

_____

3. How does camouflage help an organism survive?

_____

_____

Inquiry Activity
Color and Heat

You will investigate how the color of an animal's body covering can affect its temperature.

Write a Hypothesis Will a darker or lighter colored covering keep an animal warmer in the Sun?

If a body covering is a darker color, then...

_____

_____

Materials
☐ 2 clear plastic cups
☐ 1 cup of white beans
☐ 1 cup of black beans
☐ 2 thermometers

Carry Out an Investigation

1. Place the white beans in one of the cups. Place the black beans in the other cup.

2. Place a thermometer in each of the cups.

3. Place both cups in a sunny window.

4. **Record Data** Measure the temperature of each cup every hour throughout the day. Record your data in the table below.

Temperature (°C)		
Time	**Black Beans**	**White Beans**

Communicate Information

4. Did the data support your hypothesis? Explain.

_____

_____

_____

Desert Adaptations

📖 Read pages 96–97 in the *Science Handbook.* Answer
   the following questions after you have finished reading.

5. What adaptations help desert plants live for long periods
   of time without water?

_____

_____

_____

_____

⚙️ Crosscutting Concepts
Cause and Effect

6. Which colors of body coverings help desert animals
   survive in their environment? Why?

_____

_____

_____

Forest Adaptions

📖 Read pages 98–99 in the *Science Handbook.* Answer the following questions after you have finished reading.

7. What are the two main types of forests? How are they similar and different?

_____

_____

_____

_____

8. How does hibernation help animals survive in a temperate forest?

_____

_____

Rabbit Population

▦ Investigate how the color of an animal can help it survive by conducting the simulation. Answer the questions after you have finished.

9. Which color of rabbits survived better in the desert environment? In the snowy environment? Use evidence from the simulation to help you explain why.

_____

_____

_____

_____

_____

_____

Ocean and Wetland Adaptions

📖 Read pages 100–102 in the *Science Handbook.* Answer the following questions after you have finished reading.

10. How do air bladders help algae in the ocean survive?

_____

_____

_____

11. What types of adaptations do ocean animals have?

_____

_____

_____

12. Why do animals migrate?

_____

_____

_____

13. How have wetland plants and animals adapted to survive in this environment?

_____

_____

_____

_____

_____

Inquiry Activity
Animal Fat

You will investigate how fat helps keep animals warm in cold environments.

Write a Hypothesis Can fat help keep your finger warm in cold water?

If my finger has a layer of fat, then, when placed in ice water...

_____

_____

Materials

☐ paper towel

☐ vegetable fat

☐ bowl

☐ ice water

☐ stopwatch

Carry Out an Investigation

1. Use a paper towel to spread vegetable fat over one index finger. Try to coat your finger completely. Leave your other index finger uncovered.

2. Put one index finger into the ice water as your partner starts the stopwatch. Have your partner time how long you can keep your finger in the water. Repeat with your other finger. Record your data on a separate sheet of paper.

3. Trade roles with your partner and repeat steps 1 and 2.

Communicate Information

14. Did the results support your hypothesis? How might fat help animals survive in cold ocean environments?

_____

_____

_____

FOLDABLES

Cut out the Notebook Foldables tabs given to you by your teacher. Glue the anchor tabs as shown below. Use what you have learned to explain how each adaptation helps animals survive. Give an example of each adaptation.

Glue anchor tab here

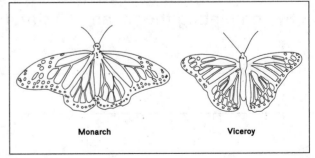

Monarch   Viceroy

15. Draw an example of an animal and their adaptations. Label the adaptations.

Science and Engineering Practices

Use examples from the lesson to explain what you can do!

Think about how you used the data that you collected throughout the lesson. Tell how you can engage in argument from evidence by completing the "I can . . ." statement below.

I can _____

_____

_____

Research, Investigate, and Communicate

Inquiry Activity
Minnow Observations

Materials

☐ minnow tank from Lesson 1

You will observe the minnows from Lesson 1 to observe their adaptations to a water environment. Think about what minnows look like. Find out about their typical habitat.

Make a Prediction What types of adaptations will minnows have?

_____

_____

Carry Out an Investigation

1. Observe the minnow tank that you set up in Lesson 1. On a separate sheet of paper, record any adaptations the fish have to their typical environment.

2. Look at the fish from the top, from the bottom, and from each side of the tank. Observe how they would blend in with their typical environment.

Communicate Information

1. **Make an Argument** What color are the fish? Are they the same color on the top, bottom, and sides? How does this help them survive in their habitat?

_____

_____

_____

Glue your observations here.

Performance Task
Design a Bird

As an ornithologist, you will design a model bird that is well-adapted to its environment. You will use evidence to explain how the adaptations will help it survive.

Ask a Question What question will your research help you to answer?

_____

_____

Materials
☐ various classroom resources: cardboard boxes, scissors, tape, glue, straws, paper clips, cotton balls

Make a Model

1. Describe the environment where your model bird will live. Describe what the bird eats and where it finds its food.

_____

_____

_____

2. Make a sketch of your model bird below. Label its adaptations and identify the materials that you will use to build it.

3. Use classroom resources to construct your model.

Communicate Information

1. **Make an Argument** How do adaptations help your bird survive in its environment? Use evidence from the lesson.

_____

_____

_____

? Essential Question
How do adaptations help plants and animals survive?

Think about the bird photos at the beginning of the lesson. Use the birds as examples to explain how adaptations help plants and animals survive.

_____

_____

_____

_____

⚙ Science and Engineering Practices

Review the "I can . . ." statement you wrote earlier in the lesson. Explain what you have accomplished in this lesson by completing the "I did . . ." statement.

Now that you're done with the lesson, share what you did!

I did _____

_____

_____

Natural Selection

PAGE KEELEY
SCIENCE
PROBES

Will the Animals Survive?

There is a type of animal that lives in forests. They eat tree nuts and seeds. They find shelter high up in the treetops. Their fur helps them blend in with the bark of the tree. What would happen to these animals if all the trees in the forest were cut down? Circle the answer that best matches your thinking.

A. *They would all die.*

B. *Some would survive if they were born with a characteristic that would help them live in the changed environment.*

C. *Some would survive by changing their food, shelter, or fur so they could live in the changed environment.*

Explain your thinking. Describe what happens to organisms when their environment changes.

_____

_____

_____

Science in Our World

▶ Watch the video about Galápagos finches.
What questions do you have?

_____

_____

_____

_____

Read about a zoologist, and answer
the questions on the next page.

> A zoologist needs to understand how variations among organisms provide survival advantages to some.

STEM Career Connection
Zoologist

As a zoologist at the San Diego Zoo, I get
to study many different things about a lot of
different animals. My days at the zoo are never
boring. That's for sure! I am often out in the zoo
observing the animals and their behaviors, which
helps me learn about them. I take notes and
recordings of each animal. Sometimes I work in
a lab, comparing my notes and doing research. I also
give presentations to visitors at the zoo.

As a zoologist, I need to think about what
an animal needs to survive in the environment.
Most animals have specific traits that help them to
survive in the environment that they live in. I study what
those traits are and how they support the animal to find
food, to stay safe, or to benefit them in other ways.

POPPY
Park Ranger

1. What are some of the zoologist's duties at the zoo?

_____

_____

_____

_____

2. What types of things do you think the zoologist takes notes about?

_____

_____

❓ Essential Question
How do variations in traits provide advantages for survival?

_____

_____

_____

_____

⚙️ Science and Engineering Practices

I will construct explanations.

> Like a zoologist, you will use evidence to explain what you learn in this lesson.

Inquiry Activity
Giraffe Feeding

Who will get the most resources? You will use your observations of people and their individual traits to construct an argument about why some students were more successful at reaching sticky notes at various heights on a wall.

Make a Prediction Which students will be more successful at collecting the sticky notes?

_____

_____

_____

<table>
<tr><td>**Materials**</td></tr>
<tr><td>☐ sticky notes</td></tr>
<tr><td>☐ stopwatch or clock with a second hand</td></tr>
</table>

Carry Out an Investigation

1. You will have one minute to gather as many sticky notes (leaves) from the walls as possible. You are working alone and can only use your hands. You may not stand on anything other than the floor. Start when your teacher says "go."

2. **Record Data** Count the number of leaves you collected. Record the numbers in the table below.

3. Share your data with some classmates. Gather and record data from your classmates.

Student	1	2	3	4	5
Number of Leaves Collected					

Communicate Information

1. **Analyze Data** What traits do you notice about the students who were able to gather the most leaves? What characteristics do they have?

_____

2. How did their characteristics help them gather more leaves?

_____

_____

3. If the sticky notes represent leaves, would you expect similar results if you were to observe giraffes in the wild? Explain.

_____

_____

_____

4. **Construct an Explanation** Use your observations to explain about how differences in characteristics lead to survival advantages in the wild.

_____

_____

_____

_____

Obtain and Communicate Information

Vocabulary

Use these words when explaining natural selection.

predator   natural selection   competition

Trait Variations and Survival

Read *Trait Variations and Survival* on how variations in traits can allow some organisms to survive better than others. Answer the following questions after you have finished reading.

1. What is a variation?

_____

_____

2. How do trait variations affect survival?

_____

_____

_____

3. How does a better sense of smell help a deer?

_____

_____

_____

_____

Inquiry Activity
Camouflage from Predators

You will model an animal hunting. You will use your observations to construct an argument about which animal color was harder to identify.

Write a Hypothesis If an animal is the same color as its environment, then . . .

_____

_____

Carry Out an Investigation

BE CAREFUL Use caution with scissors.

1. Cut 20 small circles out of one of the sheets of yellow paper.

2. Cut 20 small circles out of a sheet of brown paper.

3. Spread out all 40 circles onto the second sheet of yellow paper.

4. Have a partner time you for 30 seconds. When your partner says, "go," pick up as many circles as you can 1 at a time from the yellow paper.

5. **Record Data** Record your data in the table below.

	Number of Yellow Circles Collected	Number of Brown Circles Collected
Me		
My Partner		

6. Switch roles with you partner and repeat steps 3–5.

Communicate Information

4. Analyze Data Did you and your partner pick up more yellow circles or more brown circles?

_____

_____

5. Construct an Explanation Which circles were camouflaged? How did this help them "survive"?

_____

_____

_____

Peppered Moths

Read *Peppered Moths* on how camouflage can help organisms survive and reproduce. Answer the following questions after you have finished reading.

6. What are the two color variations of peppered moths?

_____

_____

_____

7. How did the environment of the peppered moth change?

_____

_____

_____

_____

⚙️ Crosscutting Concepts
Cause and Effect

8. Why did the population of peppered moths change?

_____

_____

_____

_____

Variations and Natural Selection

👁 Read *Variations and Natural Selection* on examples of variations that help plants and animals survive and reproduce. Answer the following questions after you have finished reading.

9. Which variation in giraffes helped them survive? How did this change the characteristics of giraffes today?

_____

_____

_____

10. What is natural selection? Use natural selection to explain the characteristics of prickly pear thorn bushes today.

_____

_____

_____

_____

FOLDABLES

Cut out the Notebook Foldables tabs given to you by your teacher. Glue the anchor tabs as shown below. Use what you have learned to describe the different characteristics that help these turtles survive.

Glue anchor tab here

Rabbit Population

📖 Investigate how camouflage can help an animal survive by revisiting the simulation and the questions you answered on page 194 in the *Be a Scientist Notebook*. Answer the question after you have finished.

11. Use evidence from the simulation to explain how the rabbit population might change in each environment.

_____

_____

_____

_____

_____

⚙️ Science and Engineering Practices

Use examples from the lesson to explain what you can do!

Think about the observations you made and the data you collected and used to construct explanations about which organisms are more likely to survive. Tell how you can use evidence to construct explanations by completing the "I can . . ." statement below.

I can _____

_____

_____

_____

Research, Investigate, and Communicate

Inquiry Activity
Natural Selection in Minnows

Materials

☐ minnow tank from Lessons 1 and 2

You will use your minnow observations from Lessons 1 and 2 to construct an explanation about how group behaviors and adaptations help them survive.

Ask a Question What question do you hope to answer in your investigation?

_____

_____

_____

Carry Out an Investigation

1. Observe the minnow tank that you set up in Lesson 1. On a separate sheet of paper, take notes about how the fish survive in their environment and what might make some fish more able to survive than others.

2. Review your observations from Lessons 1 and 2 on page 182 and page 199 in your *Be a Scientist Notebook*.

Communicate Information

1. **Make an Argument** Is a fish that is in a group or a fish that is alone more likely to survive? Use this as evidence to explain how group behaviors have been selected.

_____

_____

_____

Glue your notes here.

2. Draw a model of a minnow showing the various colors.

3. **Construct an Explanation** How do the colors of minnows help them survive in their typical environment?

_____

_____

_____

_____

_____

_____

Performance Task
Galápagos Finches

A zoologist needs to understand how the traits of organisms help them survive in their environments. You will study the Galápagos finches and use what you learned in the lesson to explain how their adaptations help them survive.

▶ Rewatch *Variation of Traits* on the different adaptations of the Galápagos finches. Answer the questions after you have finished watching.

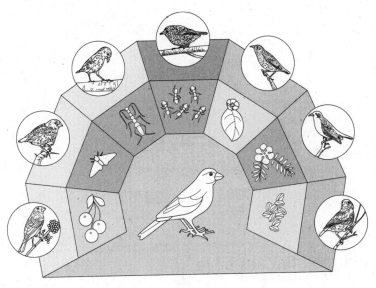

Ask a Question Write a question about the Galápagos finches that you would like to answer with your research.

_____

_____

_____

Be like a zoologist and study how the adaptations of Galápagos finches help them survive in their environment.

	Habitat	Food Source	Adaptations
Tree Finch			
Woodpecker Finch			
Warbler Finch			
Ground Finch			

1. **Construct an Explanation** Why do the finches have different adaptations?

_____

_____

_____

_____

_____

_____

_____

2. Make an Argument Would the adaptations of the ground finch help the tree finch survive? What would happen to a tree finch that was gray?

_____

_____

?Essential Question
How do variations in traits provide advantages for survival?

▷ Think about the video of Galápagos finches. Use the finches to explain how variations in traits provide advantages for survival.

_____

_____

_____

⚙ Science and Engineering Practices

Now that you are done with the lesson, share what you did!

Review the "I can . . ." statement you wrote earlier in the lesson. Explain what you have accomplished in this lesson by completing the "I did . . ." statement.

I did _____

_____

_____

Survival

⚙ Performance Project
Honeybee Research

Look back at your questions about honeybees from the beginning of the lesson. Beekeepers have learned a lot about bees and how they survive. They know what bees eat, how they work together, and how they make honey. What can you learn about these little animals?

Use print or media resources to research information about bees and how they work in groups to survive. Find out how bees communicate, gather food, and defend their hive. What different roles do bees have in a colony? What are the benefits of honeybees working in groups? Record your research notes on the lines below.

_____

_____

_____

_____

_____

_____

What information can you find about bees and how they survive?

Now that you have researched and learned more about bees, you are going to present your information to the class. Create a poster, brochure, news article, or journal entry to communicate your findings. Support the argument that honeybees must work in groups to survive. Use the space below to begin gathering the main ideas for your presentation.

Explore More in Our World

Are there any questions that you still have at the end of this module? If so, use your skill of gathering evidence to find answers to your questions.

Changes in Ecosystems

 Science in Our World

Study the picture of the dam. On one side is a huge reservoir for storing water. The other side has a controlled flow of water. Share what you observe. What questions do you have about it?

_____

_____

_____

_____

_____

abc Key Vocabulary

Look and listen for these words as you learn about the changes in ecosystems.

ecosystem	food chain	food web
migrate	natural hazard	organism
pollution		

How do structural engineers design structures to help keep us safe?

HANNAH
Welder

STEM Career Connection
Structural Engineer

Last week, the plans for the new dam arrived from the architect. I will review the plans to make sure that they are safe for the people and environment near the dam. I must make sure that the design is sturdy enough to allow cars and trucks to drive across it. A flood could destroy an endangered owl's habitat downstream, so I will check that the dam is strong enough to hold back the water from the river. The architect and I will present the plan to the government next month.

How do you think a structural engineer designs strong structures?

_____

_____

_____

_____

Science and Engineering Practices

I will engage in argument from evidence.

Name _____ Date _____

Changes Affect Living Things

Changes in Ecosystems

Changes in ecosystems affect organisms. Some of these changes include the temperature, food and water supply, and shelter an organism needs to survive. Put an X in any of the boxes that best describe what can happen to a group of organisms when there is a forest fire.

None of the organisms will survive and reproduce.	Some organisms will survive and reproduce.	All the organisms will survive and reproduce.
None of the organisms will move to new locations.	Some of the organisms will move to new locations.	All of the organisms will move to new locations.
None of the organisms will move into the changed environment.	Some of the organisms will move into the changed environment.	All of the organisms will move into the changed environment.
None of the organisms will die.	Some of the organisms will die.	All of the organisms will die.

Explain your thinking. How did you decide what happens to organisms when there is a forest fire?

_____

_____

_____

Science in Our World

Look at the photos of the beaver dam. What questions do you have?

_____

_____

_____

Read about a park ranger and answer the questions on the next page.

> A park ranger pays attention to the changes in an ecosystem to see the effects on the plants and animals.

STEM Career Connection
Park Ranger

When I go to work every day, I go to a forest! I work in a national park. My job is to take care of this environment. I take care of all the living things, including the people who visit the park. I have an office and a computer, but I spend most of my day outside. I give wildlife tours and keep the trails in good shape. I check to make sure the trees and animals in the park are healthy and that the people are safe.

Two years ago, we had a small wildfire in the park. Many trees died or were injured. Now there are many new plants growing where the trees once were and many animals coming to nibble on the new plants. Deer and other animals ate the plants so fast that nothing could grow. We found a solution: We put up a fence for one year so the plants could grow big enough to survive. Then we took down the fence. Park rangers help find solutions to a lot of problems!

POPPY
Park Ranger

1. What living things does a park ranger take care of?

_____

_____

_____

2. What does a park ranger need to know to do his or her job?

_____

_____

_____

? Essential Question
How do changes in the ecosystem affect the things that live there?

_____

_____

_____

Science and Engineering Practices

I will engage in argument from evidence.

Like a park ranger, you will engage in argument from evidence about changes in ecosystems.

 Inquiry Activity
Acid Rain

How might air pollutants that change an ecosystem affect the things that live there? Acid rain is when air pollutants fall to the ground as precipitation.

Make a Prediction How might acid rain affect the growth of plants?

_____

_____

Materials
☐ safety goggles
☐ graduated cylinder
☐ marker
☐ white vinegar
☐ water
☐ 2 spray bottles
☐ 2 bean plants
☐ ruler

Carry Out an Investigation

BE CAREFUL Wear safety goggles when using liquids.

1. Make acid rain by mixing 10 mL of vinegar with 1 L of water. Pour mixture into spray bottle. Label the spray bottle Acid Rain.

2. Pour 1 L of water into the other spray bottle. Label the spray bottle Water.

3. Label one bean plant Acid Rain. Measure the height of the bean plant. Observe the plant's color and leaves. Record in the data table on the next page.

4. Label one bean plant Water. Measure the height of the bean plant. Observe the plant's color and leaves. Record in the data table on the next page.

5. Spray the Acid Rain plant with the acid rain solution until the leaves and soil are wet. Spray the Water plant similarly with the water solution. Do this every other day for 2 weeks.

6. **Record Observations** At the end of week 1 and week 2, measure the height of the bean plants. Observe the plants' colors and leaves.

Effects of Acid Rain

		Day 1	End of Week 1	End of Week 2
Plant labeled: Water	Measurement (cm)			
	Observations			
Plant labeled: Acid Rain	Measurement (cm)			
	Observations			

7 **Analyze data** Use the data to create a bar graph to compare the growth of each bean plant.

Communicate Information

1. Describe your graph.

_____

_____

_____

2. What evidence did you find that the vinegar (acid rain) affected plant growth?

_____

_____

_____

_____

Glue graph here.

Obtain and Communicate Information

Vocabulary

Use these words when explaining how changes in environments affect living things.

pollution	population	ecosystem
migrate	organism	accommodation
food chain	food web	

Patterns for Survival

▶ Watch *Patterns for Survival* on animals and how they survive. Answer the questions after you have finished watching.

1. Which population keeps the grass short on the prairie?

_____

_____

Crosscutting Concepts
Cause and Effect

2. What would be one effect of prairie dogs disappearing?

_____

_____

_____

Ecosystems

📖 Read pages 78–81 in the *Science Handbook.* Answer the following questions after you have finished reading.

3. What is an ecosystem?

_____

_____

4. Choose one ecosystem from the reading. What are some living and nonliving things in that ecosystem?

_____

_____

_____

_____

Food Chains and Food Webs

📖 Read pages 88–91 in the *Science Handbook.* Answer the following questions after you have finished reading.

5. Where does the first organism in most food chains get energy?

_____

6. How do the other organisms in a food chain get energy?

_____

_____

_____

_____

⚙ Crosscutting Concepts
Systems and Systems Models

7. On pages 90–91 in the *Science Handbook,* there is a model of a system called a food web. Look at the model. What does the turtle eat, and what eats the turtle?

_____

_____

Changing Ecosystems

▦ Investigate how rabbits, hawks, and grass depend on each other by conducting the simulation. Answer the following questions after you have finished.

8. How do you think the grass, rabbits, and hawks depend on each other?

_____

_____

_____

9. How does changing the number of rabbits affect the hawk population?

_____

_____

10. How might grass affected by disease affect the rabbit and hawk populations?

_____

_____

_____

Changes in Ecosystems

📖 Read pages 92–93 in the *Science Handbook.* Answer
the following questions after you have finished reading.

11. What is an example of a small change to an ecosystem?
What is an example of a larger change?

_____

_____

_____

Environments Change

Explore *Environments Change* on how changes in
the environment affect living things. Answer the following
questions after you have finished.

12. How is a living thing helped by one of these changes?

_____

_____

_____

13. Which of the changes to the environment are likely to
occur every year?

_____

_____

_____

_____

FOLDABLES

Cut out the Notebook Foldables tabs given to you by your teacher. Glue the anchor tabs as shown below. Use what you have learned throughout the lesson to describe the changes made to the ecosystems below.

Glue anchor tab here

How Humans Change Environments

Explore *How Humans Change Environments* on the kinds of changes people make, and how they affect living things. Answer the following question after you have finished.

14. Does building a road affect wildlife? If it does, is the effect good or bad? Use evidence to support your answer.

_____

_____

_____

_____

Science and Engineering Practices

Use examples from the lesson to explain what you can do!

Think about how you engaged in arguments using evidence about changes to the environment. Tell how you can engage in argument from evidence by completing the "I can ..." statement below.

I can _____

_____

_____

🔍 Research, Investigate, and Communicate

Invasive Species

👁 Read *Invasive Species* on the problem of bringing plants and animals to new ecosystems. Answer the questions after you have finished reading.

1. Why did the axis deer population rise so quickly?

_____

_____

2. What is a *native species*?

_____

_____

3. How do the axis deer change the food web in Hawaii?

_____

_____

4. What does this article tell you about bringing plants and animals to new ecosystems?

_____

_____

_____

Performance Task
Beaver Dam Pros and Cons

Think about the beaver dam at the beginning of this lesson. The dam changes the environment that affects living things. Use evidence to decide whether the dam should be removed, changed, or left alone.

Ask a Question Ask a question that you will answer to decide what to do with the dam.

_____

_____

Design a Solution

1. Consider the living things in or near the stream.

2. Conduct research to find out how beaver dams affect living things in an environment.

3. List at least three living things in a stream and describe how they are affected by the dam.

_____

_____

_____

_____

_____

4. Should park rangers remove the dam, change it, or leave it alone? On a separate sheet of paper, describe what you would do and use evidence to support it.

? Essential Question
How do changes in the ecosystem affect the things that live there?

Look at the photo of the beaver dam. Use the beaver dam as an example of something that might change an ecosystem. How does the dam affect animals and plants in, or near, the stream?

_____

_____

_____

_____

_____

Science and Engineering Practices

Now that you're done with the lesson, share what you did!

Review the "I can . . ." statement you wrote earlier in the lesson. Explain what you have accomplished in this lesson by completing the "I did . . ." statement.

I did _____

_____

_____

_____

Natural Hazards Change Environments

PAGE KEELEY
SCIENCE
PROBES

Habitat Hazards

Four friends were talking about changes to habitats that can harm the organisms that live there. They each had different ideas about what causes the harmful changes. This is what they said:

Sofia: I think harmful changes to habitats are caused by hazards created by humans.

Dean: I think harmful changes to habitats are caused by natural hazards.

Hoda: I think harmful changes to habitats are caused by both hazards created by humans and natural hazards.

Alfie: I think harmful changes to habitats are caused by hazards created by humans. Natural hazards cause changes to habitats but they are not harmful.

Who do you agree with the most? _____

Explain why you agree.

_____

_____

_____

Science in Our World

▶ Watch the video of a wildfire. What questions do you have?

_____

_____

_____

_____

_____

_____

Read about an ecologist and answer the questions on the next page.

An ecologist studies how plants and animals interact with their environment.

STEM Career Connection
Ecologist

It's been raining a lot around here, much more than usual. As an ecologist, I look at how the environment affects all the plants and animals that live here. Any change to the environment affects living things, in ways that can be hard to predict. Some plants do better with more rain, for example. Then the animals that eat these plants will have more food.

Sometimes there are big changes to the environment, which means big changes for living things. This past spring we had a tornado that knocked down a lot of trees! The smaller plants will now get more sunlight to grow. Ecologists always have a lot of changes to study.

FINN
Construction Manager

1. What do ecologists study in an environment?

_____

_____

2. What does an ecologist need to know to do his or her job?

_____

_____

_____

? Essential Question

What are natural hazards and how can they change environments?

_____

_____

⚙ Science and Engineering Practices

I will engage in argument from evidence.

Like an ecologist, you will engage in argument from evidence.

Inquiry Activity
Floods Affect Plants

When an area receives more rain than usual, a flood can occur. A flood can affect the living things in an area. What happens to plants when there is a flood?

Make a Prediction What will happen to a plant that gets too much water?

_____

_____

_____

Materials

- ☐ 3 pansy plants of the same size
- ☐ graduated cylinder
- ☐ ruler

Carry Out an Investigation

1. Label three plants A, B, and C. Water plant A every day with 30 mL of water. Water plant B every day with 60 mL of water. Water plant C every day with 120 mL of water.

2. Place the plants side by side in a sunny place, so they get the same amount of light.

3. **Record Observations** Observe changes in your plants every few days. How do the leaves look? How tall is the plant? What else do you notice?

Plant	Observations			
	1	2	3	4
A				
B				
C				

Communicate Information

1. Draw a picture of the three plants after 2 weeks.

2. Which plant was the least healthy?

_____

_____

3. What evidence could you use to argue about which plant
   is the least healthy?

_____

_____

_____

4. **Construct an Explanation** How can a flood affect plants?

_____

_____

Obtain and Communicate Information

abc Vocabulary

Use these words when explaining how natural hazards change environments.

| environment | natural hazard | flood |
| earthquake | landslide | tornado |

Environmental Changes

▶ Watch *Environmental Changes* to see what happens to environments. Answer the questions after you have finished watching.

1. What are three things that change environments?

_____

_____

2. What do you think is the next thing that will happen to the forest after a fire?

_____

_____

Crosscutting Concepts
Cause and Effect

3. What might happen if there is a long period of no rain?

_____

_____

Volcanoes and Earthquakes

📖 Read pages 144–146 in the *Science Handbook.* Answer the following questions after you have finished reading.

4. Which different ways can a volcano change land?

_____

_____

5. What causes damage from an earthquake?

_____

_____

_____

Wildfires

📂 Explore *Wildfires* on the effects of fire in a forest. Answer the following question after you have finished reading.

6. Describe one negative effect and one positive effect of a wildfire.

_____

_____

_____

Floods and Landslides

📖 Read pages 148–149 in the *Science Handbook.* Answer the following questions after you have finished reading.

7. How might land have changed after a flood dries up?

_____

_____

_____

_____

8. Natural hazards might cause other hazards to occur. What hazard might cause a landslide?

_____

_____

_____

Droughts, Landslides, Floods

🔎 Explore *Droughts, Landslides, Floods* on the effects of three natural hazards. Answer the following question after you have finished reading.

9. Compare the effects of a landslide and a flood. How are they the same?

_____

_____

_____

_____

_____

Inquiry Activity
Landslide

You will create a model of a landslide. Landslides occur mostly in hilly areas. They are usually triggered by an earthquake or flash flood.

Make a Prediction What would happen if a large portion of land moved over an area of homes?

_____

_____

Materials

- [ ] paint tray liner
- [ ] 2 cups of sand
- [ ] gram cubes
- [ ] water
- [ ] container
- [ ] 1 book

Make a Model

1. Place 1 book under the top of the paint tray liner.

2. Use 2 cups of dry sand. Place it along the top of the liner to 2 cm deep.

3. Place a small cube every 2 cm along the sand to represent houses.

4. Slowly pour water out of the container onto the sand. Observe how the flooding affects the sand. Tap the bottom of the tray to model an earthquake.

Communicate Results

10. Describe what happened to the sand.

_____

_____

11. What was the effect on the homes?

_____

_____

Weather Events—Tornadoes and Hurricanes

📖 Read pages 170 and 172 in the **Science Handbook.** Answer the following questions after you have finished reading.

12. What kind of weather does a hurricane bring?

_____

_____

_____

13. What are some ways to stay safe in hurricanes and tornadoes?

_____

_____

_____

Tornadoes and Hurricanes

🔊 Explore *Tornadoes and Hurricanes.* Answer the following question after you have finished reading.

14. How are hurricanes and tornadoes the same and different?

_____

_____

_____

Inquiry Activity
Sudden Movement

Materials
- [ ] tin pan
- [ ] sand
- [ ] blocks
- [ ] twigs

You will model an earthquake.

Make a Prediction What will happen to land after it moves suddenly?

_____

_____

Make a Model

1. Fill a pan halfway with sand. Make a mountain in the sand.

2. Place blocks in the sand to model houses. Place twigs to model trees.

3. Tap the pan gently. Observe. Now tap it harder. Observe what happens to the sand, houses, and trees.

Communicate Information

15. How can sudden movement change land?

_____

_____

16. How did tapping harder affect the results?

_____

_____

17. In this activity, you made a model for a system. What system did you model?

_____

_____

FOLDABLES

Cut out the Notebook Foldables tabs given to you by your
teacher. Glue the anchor tabs as shown below. Use what
you have learned throughout the lesson to describe
the vocabulary words using the pictures below.

Glue anchor tab here

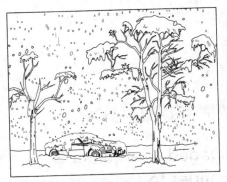
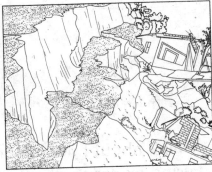

Changing Ecosystem

Investigate *Changing Ecosystem* to see how different changes affect an ecosystem. Answer the questions after you have finished.

18. How does the amount of grass affect the rabbit and hawk populations?

_____

_____

19. What evidence do you have that shows the drought affected the ecosystem?

_____

_____

_____

20. What happens to an ecosystem after a fire?

_____

_____

Science and Engineering Practices

Use examples from the lesson to explain what you can do!

Think about how you engaged in arguments using evidence about changes to the environment. Tell how you can engage in argument from evidence by completing the "I can ..." statement below.

I can _____

_____

_____

🔍 Research, Investigate, and Communicate
Natural Hazards

▶ Watch *Natural Hazards* on the various weather events. Answer the questions after you have finished watching.

1. When do we call something a natural disaster?

_____

_____

2. **Research** Find out about a natural disaster that has occurred. What type of natural disaster was it?

_____

_____

3. What evidence can you use to argue that it was a natural disaster?

_____

_____

_____

_____

Performance Task
A Wildfire Solution

Think about the photos you saw about how environments change after fires. Use evidence and engage in argument over whether small fires should be used to prevent larger fires.

Ask a Question Ask a question that you will answer to decide whether small fires should be used to prevent larger ones.

_____

_____

Design a Solution

1. Conduct research on the ways that controlled fires are being used to prevent larger fires.

2. List the positive and negative effects of fire on an ecosystem.

_____

_____

_____

3. Could a wildfire be a solution? Choose an answer and use evidence to support it.

_____

_____

_____

_____

_____

? Essential Question
What are natural hazards and how can they change environments?

▶ Rewatch the video of the wildfire. What are natural hazards, and how can they change environments?

_____

_____

_____

_____

_____

⚙ Science and Engineering Practices

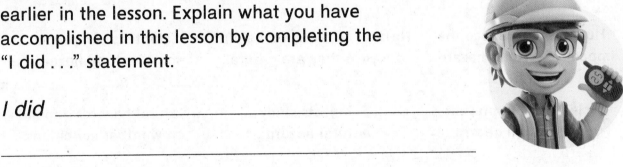

Now that you're done with the lesson, share what you did!

Review the "I can . . ." statement you wrote earlier in the lesson. Explain what you have accomplished in this lesson by completing the "I did . . ." statement.

I did _____

_____

_____

Humans and Natural Hazards

PAGE KEELEY
SCIENCE
PROBES

Natural Hazards

Natural hazards impact how humans and other organisms live. Put an X in any of the boxes that best describe natural hazards.

Natural hazards can result from natural processes.	Humans cause most natural hazards.	Humans can stop most natural hazards.
Humans can reduce the impact of natural hazards.	Natural hazards are helpful because they are natural.	Scientists can predict some natural hazards.
Scientists can prevent most natural hazards.	Scientists study natural hazards.	Natural hazards depend on weather conditions.

Explain your thinking. Describe your ideas about natural hazards.

_____

_____

_____

_____

🌐 Science in Our World

▶ Watch the video of the storm surge. Observe the wind and the rain. What questions do you have?

_____

_____

_____

_____

Read about a civil engineer and answer the questions on the next page.

Civil engineers must understand the effects of natural hazards to build safe structures.

STEM Career Connection
Civil Engineer

As a civil engineer, I design and oversee large construction projects for my city. I specialize in bridge design. I must make sure that the bridges that I design are safe. That might not seem like a big deal, but I live in an area that is at a high risk for earthquakes!

I must make sure that the bridges I design will stand up to strong earthquakes. This takes a lot of planning, many tests, and careful selection of materials. Often, our designs include frames that have braces to help hold them up, even if the ground is moving. After our design has been tested and finalized, I oversee the construction of the bridge to make sure that the plans are carried out correctly and that everyone can cross it safely. The next time you use a bridge, think of a civil engineer!

FINN
Construction Manager

1. What do civil engineers do?

_____

_____

_____

2. What process does the engineer follow to design bridges?

_____

_____

_____

_____

? Essential Question

How can humans reduce the impact of natural hazards?

_____

_____

_____

⚙ Science and Engineering Practices

I will engage in argument from evidence.

Like a civil engineer, you will use evidence to make a claim about a solution to a design problem.

Inquiry Activity
Building Sugar Structures

Materials

☐ book

☐ 20 sugar cubes

Can a model building withstand an earthquake?

Make a Prediction How many sugar cubes can be stacked without falling over when they are tapped?

_____

_____

_____

Carry Out an Investigation

BE CAREFUL Never eat anything used for a science activity.

1. Lay a book flat on your desk. Place 1 sugar cube in the center of the book. The sugar cube represents a building.

2. Tap one edge of the book gently to simulate an earthquake. Observe the cube.

3. Stack two cubes on top of each other. Tap the book gently with the same force and observe.

4. Continue to stack the cubes one at a time. Tap the book after you add each cube. See how many cubes you can stack before the stack falls over.

Sugar Cubes Stacked	
Number of Cubes	Observations

4 **Analyze Data** If you use the same tapping force each time, what is the largest number of cubes that will remain stacked after tapping the book?

_____

_____

Communicate Information

1. How did your model building stand up to the shaking?

_____

_____

2. **Make an Argument** Use the evidence that you collected to tell how your model might show how an earthquake affects a building.

_____

_____

_____

_____

_____

Obtain and Communicate Information

Vocabulary

Use these words when explaining how humans prepare for natural hazards.

| levee | floodwall | lightning rod |

Scientists Study Natural Hazards

Read *Scientists Study Natural Hazards* on how scientists learn more about hurricanes, wildfires, earthquakes, and floods. Answer the following questions after you have finished reading.

1. What do scientists need to know about wildfires to solve problems and help people?

_____

2. How can studying the locations of earthquakes help communities?

_____

_____

_____

3. How might knowing which materials stand up to the strongest winds help engineers?

_____

_____

_____

Building Structures

Explore the Digital Interactive *Building Structures* on how buildings are designed in areas that are at a high risk for earthquakes. Answer the questions after you have finished.

4. What is the purpose of shock absorbers?

_____

_____

_____

5. How is the steel inside the building constructed differently than in other buildings?

_____

_____

_____

Humans and Natural Hazards

Watch *Humans and Natural Hazards* on how the effects of floods, tornadoes, earthquakes, hurricanes, and wildfires can be reduced. Answer the questions after you have finished watching.

6. How do scientists help us prepare for earthquakes?

_____

_____

_____

7. How can the effects of all types of natural hazards be reduced?

_____

_____

Inquiry Activity
Landslides and Sandbagging

You will return to your landslide setup from Lesson 2 and use sandbags to reduce the effect of the water on the land.

Write a Hypothesis How will the sandbags change the effect of the water on the land?

If sandbags are used to reduce the amount of water that flows over the land, then...

Materials

☐ snack bags

☐ sand

☐ string

☐ landslide setup from lesson 2

☐ water

_____

Carry Out an Investigation

1. Make your sandbags by filling each bag with 4 to 5 spoonfuls of sand. Tie the snack bags with the string to hold the sand in.

2. Replace the houses and sand on your landslide setup.

3. Build a wall out of the sandbags on your landslide setup. Position your wall so that it blocks the water from flowing over the part of your pan that has the houses.

4. Run the water over the landslide.

5. **Record Data** What did you observe?

_____

_____

Communicate Information

8. **Make an Argument** Use your observations from the activity to tell how well the wall of sandbags worked to reduce the effects of a landslide.

_____

_____

_____

FOLDABLES®

Cut out the Notebook Foldables tabs given to you by your teacher. Glue the anchor tabs as shown below. Use what you have learned to explain how humans can help prevent or lessen the damage from natural hazards.

> Glue anchor tab here

> Glue anchor tab here

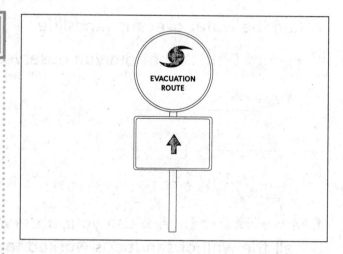

Preparedness

👁 Read *Be Prepared* on how we can be ready for natural hazards. Answer the following question after you have finished reading.

9. How can you be prepared for the natural hazard that is most likely to affect your area?

_____

_____

_____

_____

_____

⚙ ## Science and Engineering Practices

Use examples from the lesson to explain what you can do!

Think about how the data you have collected tells how well different solutions work to reduce the impacts of natural hazards. Tell how you can engage in argument from evidence by completing the "I can . . ." statement below.

I can _____

_____

_____

_____

Research, Investigate, and Communicate

Lightning Rod

Explore the Digital Interactive *Lightning Rod* on how lightning rods prevent damage caused by lightning. Answer the following questions after you have finished.

1. How do lightning rods work? Use the interactive and additional print and digital resources to research the topic. Take notes in the space below.

_____

_____

_____

_____

_____

2. **Make an Argument** How well do lightning rods work to prevent storm damage? After you have completed your research, write a report on a separate sheet of paper in which you make a claim, supported by evidence arguing how well lightning rods work to reduce the impact of storms. Summarize the argument you will make below.

_____

_____

_____

Performance Task
Building Weatherproof Structures

As a civil engineer, you will build and perform tests on a hurricane-proof model building. You will communicate and make an argument.

Define a Problem How does the number of floors affect the stability of a building during a hurricane?

_____

_____

Design a Solution

BE CAREFUL Never eat anything used for a science activity. Use caution when handling the toothpicks, as well as using a fan and water.

1. On a separate sheet of paper, draw and label a design for a sturdy building that has more than one level. It will need to withstand wind and rain, and it must be built out of up to 40 marshmallows and 40 toothpicks.

2. Carry out your plan and build your design. Construct your model building on top of the gelatin.

3. Place the container of gelatin with the model building in the large plastic tub. Turn the fan on low with the air directed at the structure for 30 seconds. Gently pour water over the structure.

4. **Record Data** Record your observations below.

_____

_____

Materials

- [ ] gelatin set in a plastic container
- [ ] 40 mini marshmallows
- [ ] 40 toothpicks
- [ ] large plastic tub
- [ ] fan
- [ ] watering can full of water
- [ ] modeling clay
- [ ] craft sticks

5 Make modifications to your model building. You may choose to use the other materials provided.

6 **Test Your Solution** Empty the plastic tub and place your new model building inside. Use the fan and the watering can to apply wind and water to the building for 30 seconds.

7 **Record Data** Record your observations below.

_____

_____

Communicate Information

1. How did your modified model building stand up to the wind and water?

_____

_____

2. **Make an Argument** Use evidence that you collected in the activity to tell how well your improved model building met the requirements that you identified and solved the problem of withstanding wind and water.

_____

_____

_____

_____

_____

3. How could your design be used to help humans reduce the impact of a storm surge?

_____

_____

_____

⚙ Crosscutting Concepts
Cause and Effect

4. How can people reduce the impacts of hurricanes?

_____

_____

_____

_____

_____

5. Besides constructing a building that can withstand a hurricane, how can people reduce the impacts of weather-related hazards?

_____

_____

_____

_____

_____

_____

_____

? Essential Question
How can humans reduce the impact of natural hazards?

▶ Think about the video of the storm surge you watched at the beginning of the lesson. How can humans reduce the impact of a storm surge?

_____

_____

_____

_____

⚙ Science and Engineering Practices

Now that you're done with the lesson, share what you did!

Review the "I can . . ." statement you wrote earlier in the lesson. Explain what you have accomplished in this lesson by completing the "I did . . ." statement.

I did _____

_____

_____

_____

_____

Changes in Ecosystems

⚙️ Performance Project
Landscaping: The Effects of Flooding on Buildings and Plants

Look back at the questions that you wrote at the beginning of the module. Now that you have learned about changes in ecosystems you can probably answer many of those questions.

Structural engineers study designs made by architects. Architects are concerned with how a structure looks and is used, but a structural engineer is concerned with the safety of a structure. They work together to revise the plans. These plans are then submitted for approval. After the plans are approved, the structure can be built. The structural engineer also examines that structure while it is being built to make sure that there are no safety issues that occur during construction.

List some ways in which the structural engineer at the beginning of the module helped to design the dam to make it safe.

_____

_____

_____

_____

Act like a structural engineer. Using what you have learned, design a model landscape that is safe from flooding.

Make a Model

1. Fill the pan half full of sand.

2. Place the blocks into the sand to represent buildings. Now place the cotton swabs or toothpicks in the sand to represent plants. Make a sketch of your model on a separate sheet of paper.

3. Create a slope by placing the book under one end of the pan.

4. Spray water on the pan. Use consistent and evenly spaced sprays. Count the number of sprays it takes until the water begins to pool on the surface of the sand. Count the number of sprays it takes to move the buildings and trees. Record your observations.

Materials
☐ rectangular cake pan
☐ sand
☐ small wood or plastic snap blocks
☐ cotton swabs or toothpicks
☐ a book
☐ spray bottle filled with water

_____

_____

_____

_____

How do structural engineers design structures to help keep us safe?

5 Think of other materials you might use to prevent the buildings and plants from being damaged. Design a way to prevent the buildings and plants from moving with the same number of sprays. Draw a diagram of the pan and show the placement of each building and plant.

6 Repeat steps 1—4. Make a claim. Did the building or plants move this time? Use evidence to engage in an argument. Were they affected by the same amount of water? Reflect on the changes you made and how they affected the buildings and plants.

_____

_____

 # Explore More in Our World

Did you learn the answers to all of your questions from the beginning of the module? If not, how could you design an experiment or conduct research to help answer them?

Learn from the Past

Science in Our World

Look at the photo of the fossil. Fossils give us clues to plant and animal life in the past. What questions do you have about the fossils?

_____

_____

_____

_____

_____

abc Key Vocabulary

Look and listen for these words as you learn about the past.

endangered	extinct	fossil
paleontologist	skeleton	

How did the fossil of
a sea creature end up
in the desert?

JIN
Paleontologist

STEM Career Connection
Paleontologist

Paleontologists are detectives. We work to uncover clues buried in the ground. The types of clues that we find provide information about animals and plants that lived long ago. Many of our clues come from fossils. A fossil can show us how and where an organism lived. We recently did a dig in the desert. To our surprise, we discovered a fossil of an ancient sea creature. This information can help us learn things about the history of this location.

What questions do you have for a paleontologist?

_____

_____

_____

_____

_____

_____

Science and
Engineering Practices

I will analyze and interpret data.

Things from Long Ago

PAGE KEELEY
SCIENCE
PROBES

Extinct Today

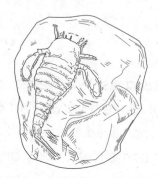

Four friends went to a science museum. They looked at the fossils of extinct plants and animals that lived millions of years ago. They had different ideas about why the plants and animals are no longer living today. This is what they said:

Norma: *I think they died out because they eventually grew too big.*

Max: *I think they just got too old and died out.*

Eshana: *I think their environment changed and they could no longer survive.*

Diego: *I think they were overhunted by humans who used them for food, clothes, and shelter.*

Which friend do you agree with the most? _____

Explain why you agree.

_____

_____

_____

Science in Our World

Look at the picture of the mastodon. What questions do you have?

_____

_____

_____

_____

Read about a museum curator and answer the questions on the next page.

> A museum curator creates displays of interesting artifacts to teach people about the past.

STEM Career Connection
Museum Curator

Every day, I work with objects that were on Earth long before human beings existed. This natural history museum has fossils from all over the world. We have dinosaur bones that are 75 million years old and a fish fossil that's 200 million years old! As a museum curator, I help get these amazing objects for the museum. I decide how to display these objects and teach people about them. Part of my job is to keep the fossils clean and safe and to fix parts that are broken. These fossils lasted millions of years underground, protected by dirt and rock. I want to make sure that they last another 200 million years!

MAYA
Geologist

1. What does this museum curator do?

_____

_____

_____

2. What helps fossils last a long time?

_____

_____

_____

? Essential Question
What happened to organisms no longer living on Earth?

_____

_____

_____

Science and Engineering Practices

I will analyze and interpret data.

Like a museum curator, you will analyze objects and teach people about life long ago.

Inquiry Activity
Model of Survival

How can I model a school of fish struggling
to survive?

Ask a Question You will make a model to show
births and deaths in a school of fish. Ask a question
that your model can answer.

_____

_____

Make a Model

1 Sturgeon are a type of fish. Count out 20 pennies to
represent a school of sturgeon.

2 Divide a piece of construction paper into six sections
to make a game board. Label the sections like this:

1. death	2. life	3. death
4. life	5. offspring	6. life

3 Carefully toss all 20 pennies onto the board.
Take away pennies that landed in sections 1 or 3.
Add a penny for any pennies in section 5.

4 Take all the pennies off the board to prepare for
another round.

5 Repeat steps 3–4 for 20 seconds. Twenty seconds
represents 20 years.

6 **Record Data** How many pennies are left? _____

7 **Interpret Data** Is your school of fish in danger of dying off? How do you know?

_____

_____

Communicate Information

1. Why did the number of pennies change?

_____

_____

2. **Construct an Explanation** What does this activity tell you about species and survival?

_____

_____

_____

3. What would happen to the sturgeon species if all sturgeon in the world did not survive?

_____

_____

4. In the activity, why did 20 rounds represent 20 years?

_____

_____

Obtain and Communicate Information

abc Vocabulary

Use these words when explaining changes to species and environments over time.

endangered   extinct

Endangered or Extinct?

👁 Read the Science File *Endangered or Extinct?* Answer the following questions after you have finished reading.

1. What is the difference between a species that is *endangered* and a species that is *extinct?*

_____

_____

2. Why might a plant or animal species become endangered?

_____

_____

⚙ Crosscutting Concepts
Scale, Proportion, and Quantity

3. Was the Model of Survival activity a good model to show how a type of organism could become endangered or extinct? Explain.

_____

_____

Extinct Animals

Explore the Digital Interactive *Extinct Animals* on species that are no longer alive. Answer the questions after you have finished.

5. How do scientists learn about extinct animals?

_____

_____

6. Which animals were alive most recently: mammoths, saber-toothed tigers, or dinosaurs?

_____

_____

Extinct and Protected Animals

Explore the Digital Interactive *Extinct and Protected Animals* to compare. Answer the questions after you have finished.

7. Why are orangutans endangered?

_____

_____

8. How are people trying to make sure elephants do not become extinct?

_____

_____

_____

FOLDABLES

Cut out the Notebook Foldables given to you by your teacher. Glue the anchor tabs as shown below. Define the words using what you have learned.

Glue anchor tab here

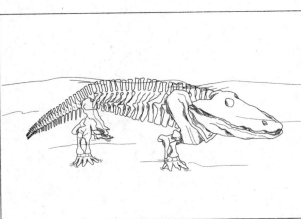

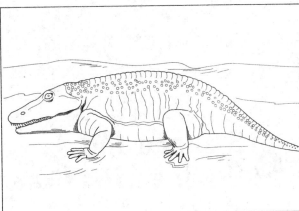

9. Draw a picture of an animal that is currently protected. Show how people are helping protect it.

Science and Engineering Practices

Use examples from the lesson to show what you can do!

Think about how you interpreted data from the *Model of Survival* activity to understand more about endangered animals. Tell how you can analyze and interpret data by completing the "I can . . ." statement below.

I can _____

_____

_____

_____

Research, Investigate, and Communicate
Horseshoe Crab Research

Horseshoe crabs are one of the oldest types of organisms on Earth. Research to learn more about why some scientists are interested in horseshoe crabs.

Research Scientists have concluded that ancestors of the horseshoe crab date back almost 450 million years.

Ask a Question What question will your research help you to answer?

_____

_____

1 Research to find out about horseshoe crabs. Take notes on what you find.

_____

_____

_____

_____

_____

Communicate Information

1. **Construct an Explanation** Why are some scientists interested in horseshoe crabs?

_____

_____

⚙ Performance Task
Research an Extinct Animal

You will research to find out more about an extinct animal.

Research Find a list of animals that are extinct. Choose an animal that became extinct in the last 200 years. Find out why the animal became extinct.

Ask a Question What question will your research help you to answer?

_____

_____

1. Conduct research to find out what changes made the animal go extinct. Write down all the reasons.

2. Share your research with your classmates.

3. **Analyze Data** What are the main causes of recent extinctions?

_____

_____

_____

_____

_____

_____

_____

_____

_____

Communicate Information

1. **Construct an Explanation** What changes caused animals to become extinct in the last 200 years?

_____

_____

_____

? Essential Question
What happened to organisms no longer living on Earth?

Think about the picture of the mastodon you saw at the beginning of the lesson. Use what you learned about endangered and extinct animals to answer the question.

_____

_____

_____

⚙ Science and Engineering Practices

Now that you're done with the lesson, share what you did!

Review the "I can . . ." statement you wrote earlier in the lesson. Explain what you have accomplished in this lesson by completing the "I did . . ." statement.

I did _____

_____

_____

Fossils

Fossil Evidence

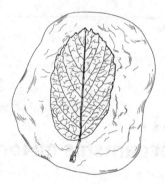

Paleontologists are scientists who study fossils for clues about the past. Put an X in any of the boxes that describe how fossils provide evidence to paleontologists about the past.

Fossils contain evidence of the types of organisms that lived in the past.	Fossils show evidence of the color of past organisms.	Fossils show how much an organism weighed when it died.
Fossils contain evidence of what environments were like in the past.	Fossils provide evidence of the size of an organism.	Fossils provide evidence of how past organisms are similar to some present-day organisms.
Fossils provide evidence of similarities and differences among past organisms.	Fossils provide evidence of the sounds a past organism made.	Fossils show how present-day land areas may have once been covered by ocean.

Explain your thinking. Describe your ideas about how fossils can help paleontologists understand the past.

_____

_____

_____

🌐 Science in Our World

Look at the photo of the paleontologist working.
Why might a paleontologist look at a skeleton? What
questions do you have?

_____

_____

_____

Read about a micropaleontologist and answer
the questions on the next page.

> A micropaleontologist
> studies the smallest
> organisms from
> long ago!

STEM Career Connection
Micropaleontologist

Some paleontologists study dinosaurs that
were taller than a house. I am a paleontologist who
studies fossils that are so tiny that you need a
microscope to see them! I look at tiny animals that lived
in the oceans and lakes. I might even look at prehistoric
pollen. The earliest life on Earth was small, so
micropaleontologists sometimes look at fossils that are
three *billion* years old!

I work in the field digging up fossils. Then I work
in the lab, studying the fossils I found. I look at them
under powerful microscopes. I conduct tests to find
out what chemicals are in the fossil. This tells me what
the environment was like when the organisms were alive.

Earth has changed a lot since life began, and there
are more extinct species than living ones. So I'll never
run out of things to study!

MAYA
Geologist

1. What does a micropaleontologist do?

_____

_____

2. What tools help a micropaleontologist?

_____

_____

_____

? Essential Question
What can we learn from fossils?

_____

_____

_____

⚙ Science and Engineering Practices

I will **analyze** and interpret data.

Like a micropaleontologist, you will analyze and interpret data to learn about living things from long ago.

✋ Inquiry Activity
Layers and Fossils

What do layers of Earth tell us about fossils?

Make a Prediction How do older and newer fossils form in one place?

_____

_____

Materials
- ☐ tablespoon
- ☐ white glue
- ☐ water
- ☐ paper cup
- ☐ colored sand
- ☐ "fossil" objects

Make a Model

1. Mix a tablespoon of glue with a little water in a cup.

2. Pour a thin layer of colored sand into a paper cup. Add a "fossil" object. Cover the object with sand of the same color.

3. Add some of the glue and water mixture to the sand to make it stay together.

4. Repeat steps 2 and 3 two more times so you have three layers of "soil" and "fossils."

Communicate Information

1. How is your model like Earth?

_____

_____

2. In your model, which rock layer formed first? Which fossil is the oldest?

_____

_____

_____

Obtain and Communicate Information

abc Vocabulary

Use these words when explaining how scientists study organisms and environments from long ago.

fossil　　　**paleontologist**　　**skeleton**

Fossils

📖 Read pages 174–175 in the *Science Handbook*. Answer the following questions after you have finished reading.

1. How are preserved remains different from fossils in stone?

_____

_____

_____

2. How do trace fossils form?

_____

_____

_____

3. What can scientists learn from studying a mold fossil of an ancient seashell?

_____

_____

_____

Inquiry Activity
How Fossils Are Made

Make a model of one type of fossil: a living thing trapped in tree resin, or amber. Observe the results to learn how amber can preserve living things.

Materials

☐ plastic spoon

☐ glue

☐ paper towel

☐ 2 apple slices

Make a Model

BE CAREFUL Do not eat science materials!

1. Hold a spoon over a paper towel. Squeeze a small amount of glue onto the spoon. Let the glue set for 10 minutes. The glue represents sticky tree resin.

2. Place an apple slice on top of the glue. This step represents an organism that got trapped in tree resin.

3. Slowly add more glue. Completely cover the apple slice.

4. Put the spoon on a paper towel. Label it "1." Place another apple slice next to the spoon. Label it "2."

5. Record Data Look at the apple slices throughout the day. Record the time and any changes you see.

Observations		
Time	**Slice 1**	**Slice 2**

6. Analyze Data Circle the apple slice that changed the most.

Communicate Information

4. Construct an Explanation How does being trapped in amber preserve living things?

_____

What Fossils Tell Us

📖 Read pages 176–177 in the **Science Handbook.** Answer the following questions after you have finished reading.

5. Trilobites are extinct. How do scientists know they lived in oceans throughout the world?

_____

_____

6. Paleontologists found fossils of a coral reef in the middle of the United States. What does that tell them?

_____

_____

✋ Inquiry Activity
Layers and Fossils, Part 2

You will uncover the "fossils" in the layers of sand that you made earlier in this lesson.

Materials
☐ "fossil" cups
☐ paintbrush

Make a Model

1 Trade cups with another group. Carefully peel away the paper from the cup. Starting at the top, brush away sand to uncover the fossils.

2 **Record Data** Record your findings in this table.

Layer	Fossil
Top	
Middle	
Bottom	

3 **Interpret Data** Circle the oldest fossil.

Communicate Information

7. Construct an Explanation How does this model show changes on Earth over time?

_____

_____

_____

⚙ Crosscutting Concepts
Scale, Proportion, and Quantity

8. You made a small model of a large system—Earth. How is the small model different from the large system?

_____

_____

_____

Animals of Today and Animals of Long Ago

▶ Watch *Animals of Today and Animals of Long Ago.* Answer the following questions when you are finished.

9. What are some differences between a woolly mammoth and an elephant?

_____

_____

_____

10. What animals living today are most closely related to dinosaurs?

_____

Fossils from Long Ago or Skeletons of Today

👁 Read the Science File *Fossils from Long Ago or Skeletons of Today.* Answer the following questions after you have finished reading.

11. How do scientists learn about the ways extinct animals used their bodies?

_____

_____

_____

12. How are pterodactyls similar to eagles?

_____

_____

_____

Learning from Fossils

👁 Read the Science File *Learning from Fossils.* Answer the following question after you have finished reading.

13. Scientists may find many fossils buried at the same depth. What do they know about these fossils?

_____

_____

_____

Fossil Dig

Investigate layers of a fossil dig by conducting the simulation. Answer the questions after you have finished.

14. Record Data Choose one area of the simulation and record data on the fossils you find there.

Area _____

Layer	Fossil Name	On water or land?	Observations
A			
B			
C			
D			

15. How did you know whether an organism lived on land or in water?

_____

_____

16. How did you know which fossils were older?

_____

Online Content at connectED.mcgraw-hill.com

FOLDABLES

Cut out the Notebook Foldables given to you by your teacher. Glue the anchor tabs as shown below.
Describe the fossils and animals shown in the pictures and explain what they tell us about prehistoric life.

Glue anchor tab here

Glue anchor tab here

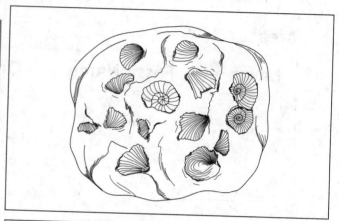

Science and Engineering Practices

Think about how you analyzed and interpreted data about fossils and where they were found, and learned more about living things and the environment long ago. Tell how you can analyze and interpret data by completing the "I can . . ." statement below.

I can _____

_____

_____

_____

_____

Use examples from the lesson to show what you can do!

Research, Investigate, and Communicate

Inquiry Activity
Fossil Mystery

Make a model of a fossil then see whether you can identify your classmate's fossil.

Make a Model

1. Choose a favorite animal. Then use the key below to draw a fossil on the next page.

If your animal is a...	then draw a...
mammal	circle
bird	square
amphibian	rectangle
reptile	triangle
fish	ball

2. Use the key below to draw marks on your fossil.

If your animal...	then mark your drawing with...
only lives in water	fins
only lives on land	feet
lives in water and on land	fins and feet
is a carnivore (eats meat)	pointed teeth
is an herbivore (eats plants)	flat teeth
is an omnivore (eats meat and plants)	pointed and flat teeth

③ **Draw your fossil below.**

④ Trade your model fossil with a person sitting next to you. Use the chart to find out about the animal your classmate chose.

Communicate Information

1. What did you learn about the fossil you investigated?

_____

_____

2. **Construct an Explanation** How do scientists use fossils to learn about extinct animals?

_____

_____

⚙ Performance Task
Tell About Animals and Environments

Use your data from the *Fossil Dig* simulation to describe animals from long ago, and their environments.

1. Describe the animals that became fossils in your area and the environments that they lived in.

A. _____

_____

B. _____

_____

C. _____

_____

D. _____

_____

2. Describe how the organisms in the different layers are different from each other.

_____

_____

_____

3. Describe how the environment changed over time.

_____

_____

_____

? Essential Question
What can we learn from fossils?

Think about the photo of the paleontologist looking at a skeleton fossil from the beginning of the lesson. Use what you learned in the lesson to explain what scientists can learn from fossils such as this.

_____

_____

_____

_____

Science and Engineering Practices

Now that you're done with the lesson, share what you did!

Review the "I can . . ." statement earlier in the lesson. Explain what you have accomplished in this lesson by completing the "I did . . ." statement.

I did _____

_____

_____

_____

_____

_____

_____

Learn from the Past

⚙ Performance Project
Looking Back

Look back at the questions that you wrote at the beginning of the module. Now that you have learned about fossils, you can probably answer many of those questions.

Paleontologists learn about organisms that lived long ago. Fossils provide clues that help the scientists gather data. They also use this information to teach others.

Think about the photo of the sea creature fossil that was found in the desert.

Research Use books or Internet sources to find information about a sea creature fossil that was found in the desert.

_____

_____

_____

_____

_____

_____

_____

What clues might a sea
creature fossil provide?

Analyze and interpret the information you found. Tell
how you think it was possible that this fossil was found
on dry land. Use module vocabulary words to write
your explanation.

_____

_____

_____

In the box below, develop a model to show how the fossil
might have formed.

┌───┐
│   │
│   │
│   │
│   │
│   │
│   │
│   │
│   │
└───┘

 ## Explore More in Our World

Did you learn the answers to all of your questions from
the beginning of the module? If not, how could you design
an experiment or conduct research to help answer them?

An Interview with

Dinah Zike Explaining
Visual Kinesthetic Vocabulary®, or VKVs®

What are VKVs and who needs them?

> VKVs are flashcards that animate words by kinesthetically focusing on their structure, use, and meaning. VKVs are beneficial not only to students learning the specialized vocabulary of a content area, but also to students learning the vocabulary of a second language.

Dinah Zike | Educational Consultant
Dinah-Might Activities, Inc. – San Antonio, Texas

Why did you invent VKVs?

> Twenty years ago, I began designing flashcards that would accomplish the same thing with academic vocabulary and cognates that Foldables® do with general information, concepts, and ideas—make them a visual, kinesthetic, and memorable experience.

I had three goals in mind:

- **Making two-dimensional flashcards three-dimensional**

- **Designing flashcards that allow one or more parts of a word or phrase to be manipulated and changed to form numerous terms based upon a commonality**

- **Using one sheet or strip of paper to make purposefully shaped flashcards that were neither glued nor stapled, but could be folded to the same height, making them easy to stack and store**

Why are VKVs important in today's classroom?

> At the beginning of this century, research and reports indicated the importance of vocabulary to overall academic achievement. This research resulted in a more comprehensive teaching of academic vocabulary and a focus on the use of cognates to help students learn a second language. Teachers know the importance of using a variety of strategies to teach vocabulary to a diverse population of students. VKVs function as one of those strategies.

Dinah Zike Explaining
Visual Kinesthetic Vocabulary®, or VKVs®

Dinah Zike's
Visual
Kinesthetic
Vocabulary®

How are VKVs used to teach content vocabulary to EL students?

" VKVs can be used to show the similarities between cognates in Spanish and English. For example, by folding and unfolding specially designed VKVs, students can experience English terms in one color and Spanish in a second color on the same flashcard while noting the similarities in their roots. "

How are VKVs used to teach content vocabulary?

" As an example, let's look at content terms based upon the combining form –vore. Within a unit of study, students might use a VKV to kinesthetically and visually interact with the terms *herbivore, carnivore,* and *omnivore.* Students note that –vore is common to all three words and it means "one that eats" meat, plants, or both depending on the root word that precedes it on the VKV. When the term *insectivore* is introduced in a classroom discussion, students have a foundation for understanding the term based upon their VKV experiences. And hopefully, if students encounter the term *frugivore* at some point in their future, they will still relate the –vore to diet, and possibly use the context of the word's use to determine it relates to a diet of fruit.

Dinah Zike's book Foldables, Notebook Foldables, & VKVs for Spelling and Vocabulary 4th-12th won a Teachers' Choice Award in 2011 for "instructional value, ease of use, quality, and innovation"; it has become a popular methods resource for teaching and learning vocabulary.

✂ cut on all dashed lines   fold on all solid lines

position

_____ is a change in

the position of an object.

1. _____ is the course

or path on which something is moving.

2. _____ is the

location of an object.

Dinah Zike's
V K V
Visual
Kinesthetic
Vocabulary®

✂ cut on all dashed lines

📋 fold on all solid lines

Memory Maker: Construct a word web. Include all words on this VKV and any other words that relate.

direc

mo

cut on all dashed lines

fold on all solid lines

compound machine

A _____ is a machine with few or no moving parts.

A _____ is two or more simple machines put together.

Dinah Zike's
**Visual
Kinesthetic
Vocabulary**®

✂ cut on all dashed lines

☐ fold on all solid lines

Memory Maker: Draw a compound machine. Label each simple machine in the compound machine.

simple

Dinah Zike's
Visual
Kinesthetic
Vocabulary®

cut on all dashed lines

fold on all solid lines

unbalanced force

are forces that act together on an object without changing its motion.

are forces that do not cancel each other out and that cause an object to change its motion.

Dinah Zike's
Visual
Kinesthetic
Vocabulary ®

✂ cut on all dashed lines   fold on all solid lines

s

Memory Maker: Use your own words to explain the difference between balanced forces and unbalanced forces.

balanced

is a push or pull.

Dinah Zike's
**Visual
Kinesthetic
Vocabulary**®

magnetic field

A ——————— is a region of magnetic force around a magnet, represented by lines.

——————— is the ability of an object to push or pull on another object that has the magnetic property.

cut on all dashed lines   fold on all solid lines

Memory Maker: The suffix -ic means "behaves like." So, magnetic means "behaves like a magnet."

How does a magnet behave, or what does it do?

_____

A _____ is an object with a magnetic force; magnets can attract or repel certain metals.

is m

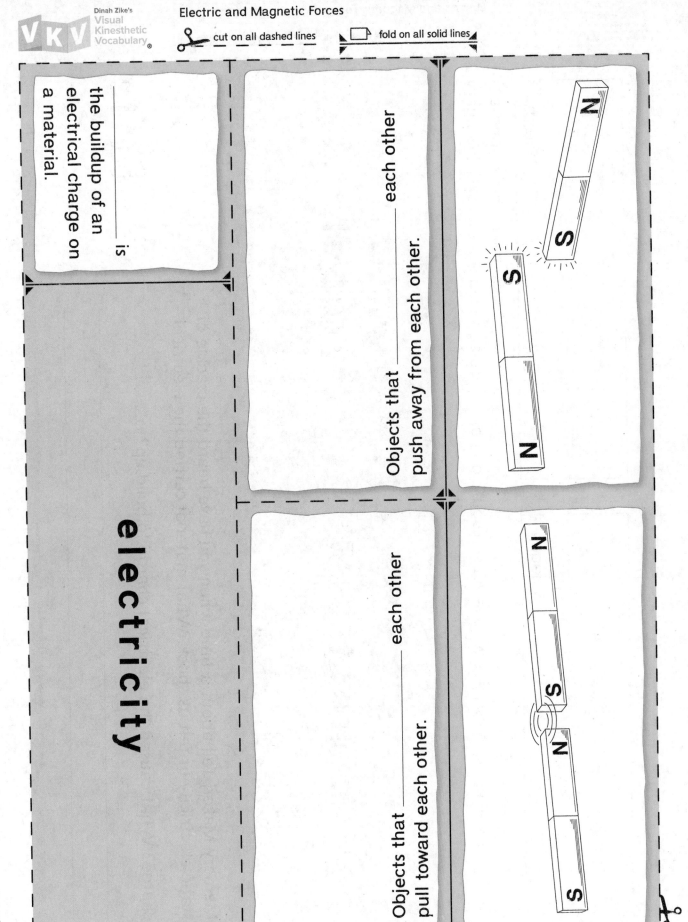

_____ is
the buildup of an
electrical charge on
a material.

electricity

Objects that
push away from each other.

_____ each other

Objects that
pull toward each other.

_____ each other

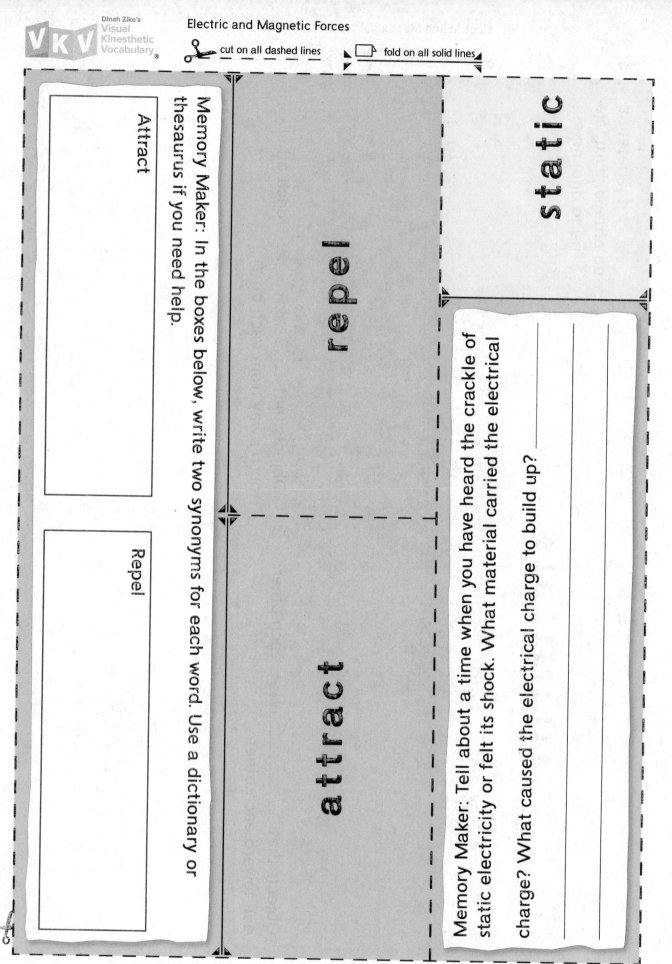

static

repel

attract

Memory Maker: In the boxes below, write two synonyms for each word. Use a dictionary or thesaurus if you need help.

Attract

Repel

Memory Maker: Tell about a time when you have heard the crackle of static electricity or felt its shock. What material carried the electrical charge? What caused the electrical charge to build up? _____

_____

_____

Dinah Zike's
Visual
Kinesthetic
Vocabulary®

✂ cut on all dashed lines

📷 fold on all solid lines

fall season

During the ____ season, the weather is hot and sometimes dry.

During the ____ season, the weather cools and becomes breezy.

A ____ is a time of the year with different weather patterns.

During the ____ season, the weather warms and becomes rainy.

During the ____ season, the weather is cold and sometimes snowy.

Dinah Zike's
Visual Kinesthetic Vocabulary ®

✂ cut on all dashed lines

▢ fold on all solid lines

Memory Maker: Draw four pictures to help you remember the weather pattern for each season.

summer

winter

spring

✂ cut on all dashed lines

fold on all solid lines

air pressure

_____ is the weight of air pressing down on Earth.

_____ is a large region of the atmosphere in which the air has similar properties.

VKV®

Dinah Zike's
Visual
Kinesthetic
Vocabulary®

✂ cut on all dashed lines

▢ fold on all solid lines

Memory Maker: In your own words, explain the difference between air mass and air pressure.

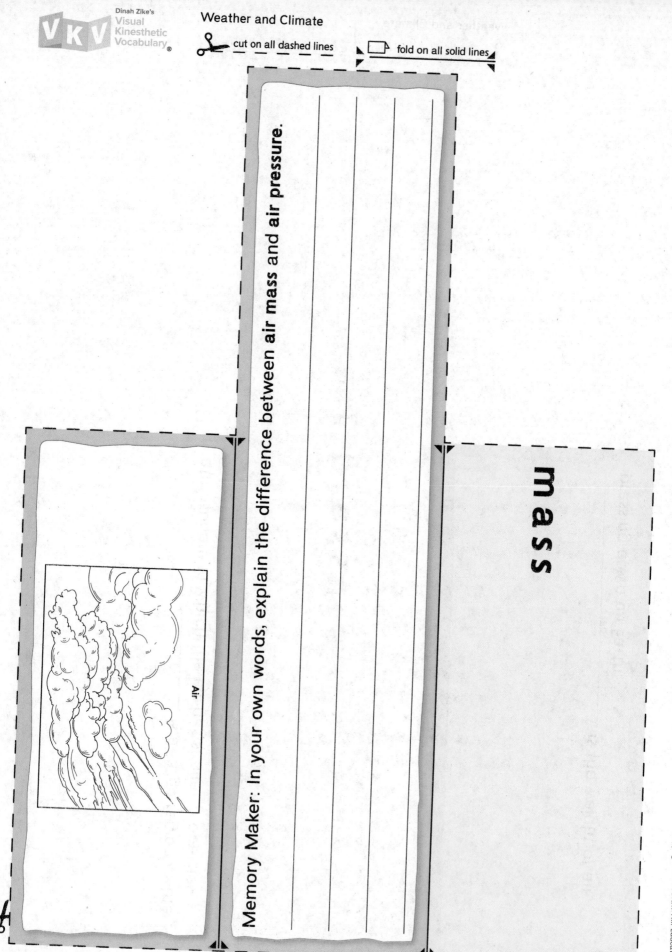

Air

mass

cut on all dashed lines

fold on all solid lines

life

_____ describes an adaptation in which an animal is active during the night and asleep during the day.

_____ means the opposite of nocturnal. Therefore, diurnal describes an adaptation in which an animal is asleep during the night and active during the day.

A _____ shows how a certain kind of organism grows and reproduces.

Dinah Zike's
Visual
Kinesthetic
Vocabulary®

Memory Maker: Below is a list of animals. Circle the names of animals that are nocturnal.

bat   squirrel   firefly   cricket   opossum   cow   bee

nocturnal

diurnal

Memory Maker: The picture on the front shows the life cycle of a plant. Next to it, draw a picture that shows the life cycle of an animal of your choice.

cycle

Dinah Zike's
Visual
Kinesthetic
Vocabulary®

✂ cut on all dashed lines

▱ fold on all solid lines

adaptation

learned behavior

An _____ adaptation is a change to a structure or behavior that helps an organism survive in its environment.

_____ is behavior that is _____ learned from watching others.

Dinah Zike's
Visual
Kinesthetic
Vocabulary®

Survival

✂ cut on all dashed lines

▭ fold on all solid lines

Memory Maker: What are some of your **learned behaviors**? For example, what are some things you do to get ready for school? Who taught you to do these things? _____

_____ is the way a person or animal acts or behaves.

Memory Maker: On the lines below, write **behavior** or **structure** to tell what the **adaptation** changed for each animal.

1. The beaks of finches are large or small depending on the finch's environment. _____

2. Monarch butterflies fly south to warmer environments during the winter. _____

Adapt is a synonym for change. Write two more synonyms for **adapt**. Use a dictionary or thesaurus if you need help.

_____

_____

Dinah Zike's
Visual
Kinesthetic
Vocabulary ®

✂ cut on all dashed lines   ▭ fold on all solid lines

To **mimic** is to look or act like someone or something else.

_____ is an adaptation in which one kind of organism looks like another kind in color and shape.

_____ means "to move from one place to another." Therefore,

_____ means "moving from one place to another."

_____ means "to rest or go into a deep sleep through the cold winter." So,

_____ means "resting or going into a deep sleep through the cold winter."

mimic

migration

hibernation

Dinah Zike's
VKV
Visual
Kinesthetic
Vocabulary ®

Survival

✂ cut on all dashed lines

▭ fold on all solid lines

e

e

ry

Memory Maker: A chameleon is an animal that is known for its **mimicry**. How does a chameleon **mimic** other organisms?

_____
_____
_____
_____

Memory Maker: Below is a list of animals. Circle the names of animals that practice **migration**.

goose   cat   salmon   butterfly   earthworm   sheep

Memory Maker: Below is a list of animals. Circle the names of animals that practice **hibernation**.

tiger   squirrel   bear   lion   chipmunk   monkey   bat

✂ cut on all dashed lines   ⬜ fold on all solid lines

produce

_____ is an organism that eats
A _____ plants or other animals.

_____ is an organism that breaks
A _____ down dead plant and animal material.

_____ is an organism, such as a
A _____ plant, that makes its own food.

Dinah Zike's
**Visual
Kinesthetic
Vocabulary**®

r

Memory Maker: Fill in the blank beside each sentence below.

1. Consumers **consume** other organisms. Write a synonym for **consume**. _____

2. Producers **produce** their own food. Write a synonym for **produce**. _____

3. Decomposers **decompose** other organisms. Write a synonym for **decompose**. _____
_____

decompos

consum

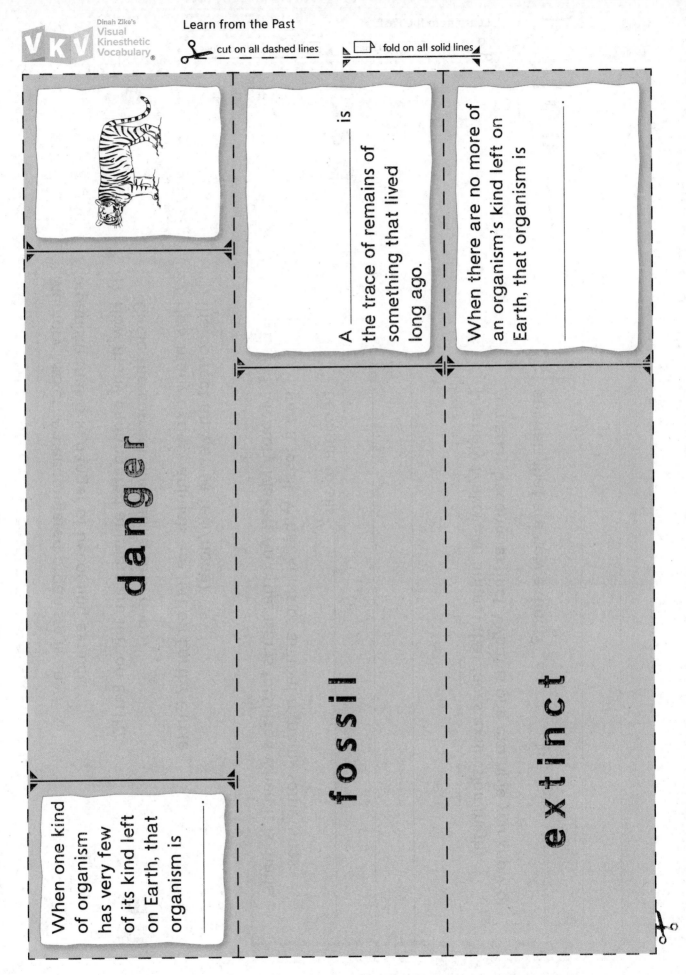

A _____ is the trace of remains of something that lived long ago.

When there are no more of an organism's kind left on Earth, that organism is _____.

When one kind of organism has very few of its kind left on Earth, that organism is _____.

danger

fossil

extinct

ion

ized

ed

Memory Maker: An endangered organism is an organism that is in danger of becoming extinct.

1) How many endangered animals are left on Earth? Circle the correct answer. (a few/none)

2) How many extinct animals are left on Earth? Circle the correct answer. (a few/none)

Memory Maker: An organism is fossilized when it is made into a fossil. What kinds of animal remains sometimes become fossils?

_____

Memory Maker: An animal that faces extinction might someday become extinct. What is one example you know of animals that are now extinct?

_____

en